The techniques of molecular biology offer a powerful means of investigating and controlling the genetic basis of mechanisms operating in living organisms. The development of these techniques in aquatic animals has now reached the stage where important questions relating to growth, development and adaptation to the environment can be addressed at the level of gene expression, and the introduction and expression of novel genes achieved. This volume presents some of the most exciting advances in this rapidly expanding area, with contributions on the evolution of adaptation to low temperature, adaptation to short-term fluctuations in temperature and salinity, gene expression during growth and development, myosin polymorphism and the generation of transgenic fish. As such, it will be of interest to all those working in the fields of marine and freshwater biology and also to those working in aquaculture.

SOCIETY FOR EXPERIMENTAL BIOLOGY
SEMINAR SERIES: 58

GENE EXPRESSION AND MANIPULATION IN AQUATIC ORGANISMS

SOCIETY FOR EXPERIMENTAL BIOLOGY SEMINAR SERIES

A series of multi-author volumes developed from seminars held by the Society for Experimental Biology. Each volume serves not only as an introductory review of a specific topic, but also introduces the reader to experimental evidence to support the theories and principles discussed, and points the way to new research.

6. Neurones without impulses: their significance for vertebrate and invertebrate systems. *Edited by A. Roberts and B.M.H. Bush*
8. Stomatal physiology. *Edited by P.C.G. Jarvis and T.A. Mansfield*
16. Gills. *Edited by D.F. Houlihan, J.C. Rankin and T.J. Shuttleworth*
31. Plant canopies: their growth, form and function. *Edited by G. Russell, B. Marshall and P.G. Jarvis*
33. Neurohormones in invertebrates. *Edited by M. Thorndyke and G. Goldsworthy*
34. Acid toxicity and aquatic animals. *Edited by R. Morris, E.W. Taylor, D.J.A. Brown and J.A. Brown*
35. Division and segregation of organelles. *Edited by S.A. Boffey and D. Lloyd*
36. Biomechanics in evolution. *Edited by J.M.V. Rayner and R.J. Wootton*
37. Techniques in comparative respiratory physiology: An experimental approach. *Edited by C.R. Bridges and P.J. Butler*
38. Herbicides and plant metabolism. *Edited by A.D. Dodge*
39. Plants under stress. *Edited by H.C. Jones, T.J. Flowers and M.B. Jones*
40. *In situ* hybridisation: application to developmental biology and medicine. *Edited by N. Harris and D.G. Wilkinson*
41. Physiological strategies for gas exchange and metabolism. *Edited by A.J. Woakes, M.K. Grieshaber and C.R. Bridges*
42. Compartmentation of plant metabolism in non-photosynthesis tissues. *Edited by M.J. Emes*
43. Plant growth: interactions with nutrition and environment. *Edited by J.R. Porter and D.W. Lawlor*
44. Feeding and texture of foods. *Edited by J.F.V. Vincent and P.J. Lillford*
45. Endocytosis, exocytosis and vesicle traffic in plants. *Edited by G.R. Hawes, J.O.D. Coleman and D.E. Evans*
46. Calcium, oxygen radicals and cellular damage. *Edited by C.J. Duncan*
47. Fruit and seed production: aspects of development, environmental physiology and ecology. *Edited by C. Marshall and J. Grace*
48. Perspectives in plant cell recognition. *Edited by J.A. Callow and J.R. Green*
49. Inducible plant proteins: their biochemistry and molecular biology. *Edited by J.L. Wray*
50. Plant organelles: compartmentation of metabolism in photosynthetic cells. *Edited by A.K. Tobin*
51. Oxygen transport in biological systems: modelling of pathways from environment to cell. *Edited by S. Eggington and H.F. Ross*
52. New insights in vertebrate kidney function. *Edited by J.A. Brown, R.J. Balment and J.C. Rankin*
53. Post-translational modifications in plants. *Edited by N.H. Battey, H.G. Dickinson and A.M. Hetherington*
54. Biomechanics and cells. *Edited by F.J. Lyall and A.J. El Haj*
55. Molecular and cellular aspects of plant reproduction. *Edited by R. Scott and A.D. Stead*
56. Amino acids and their derivatives in plants. *Edited by R.M. Wallsgrove*
57. Toxicology of aquatic pollution: physiological, cellular and molecular approaches. *Edited by E. W. Taylor*

GENE EXPRESSION AND MANIPULATION IN AQUATIC ORGANISMS

Edited by

S.J. Ennion

Department of Anatomy and Developmental Biology, Royal Free Hospital School of Medicine, London, UK
and Department of Geriatric Medicine, Royal Free Hospital School of Medicine, London, UK

G. Goldspink

Department of Anatomy and Developmental Biology, Royal Free Hospital School of Medicine, London, UK

CAMBRIDGE
UNIVERSITY PRESS

CAMBRIDGE UNIVERSITY PRESS
Cambridge, New York, Melbourne, Madrid, Cape Town, Singapore, São Paulo, Delhi

Cambridge University Press
The Edinburgh Building, Cambridge CB2 8RU, UK

Published in the United States of America by Cambridge University Press, New York

www.cambridge.org
Information on this title: www.cambridge.org/9780521570039

First published 1996
This digitally printed version 2008

A catalogue record for this publication is available from the British Library

Library of Congress Cataloguing in Publication data

Gene expression and manipulation in aquatic organisms / edited by S.J.
Ennion, G. Goldspink.
p. cm. – (Society for Experimental Biology seminar series;
58)
Includes index.
ISBN 0 521 57003 4 (hardcover)
1. Fishes – Genetics. 2. Fishes – Genetic engineering. 3. Gene
expression. I. Ennion, S. J. (Steven J.) II. Goldspink, G. III. Series:
Seminar series (Society for Experimental Biology (Great Britain)); 58.
QL638.99.G45 1996
597′.087322 – dc20 95-51195 CIP

ISBN 978-0-521-57003-9 hardback
ISBN 978-0-521-10162-2 paperback

Contents

List of contributors ix
Preface xiii

Genomic basis for antifreeze glycopeptide heterogeneity and abundance in Antarctic fishes 1
C.-H.C. CHENG

Cold-inducible gene transcription: Δ^9-desaturases and the adaptive control of membrane lipid composition 21
P.E. TIKU, A.Y. GRACEY and A.R. COSSINS

Ion transport in teleosts: identification and expression of ion transporting proteins in branchial and intestinal epithelia of the European eel 43
C.P. CUTLER, I.L. SANDERS, G. LUKE, N. HAZON and G. CRAMB

Temperature adaptation: selective expression of myosin heavy chain genes and muscle function in carp 75
G. GOLDSPINK

Crustacean genes involved in growth 93
A.J. EL HAJ

Use of the zebrafish for studies of genes involved in the control of development 113
Q. XU, K. GRIFFIN, R. PATIENT and N. HOLDER

Myosin heavy chain isogene expression in carp 123
S. ENNION

Rainbow trout myosin heavy chain polymorphism during development 149
L. GAUVRY, C. PEREZ and B. FAUCONNEAU

Transient expression of reporter genes in fish as a measure of promoter efficiency 165
N. MACLEAN, M. S. ALAM, A. IYENGAR and A. POPPLEWELL

The use of transient *lacZ* expression in fish embryos for comparative analysis of cloned regulatory elements 175
F. MÜLLER, L. GAUVRY, D.W. WILLIAMS, J. KOBOLÁK, N. MACLEAN, L. ORBAN and G. GOLDSPINK

Molecular characterization of prolactin receptor in tilapia 201
P. PRUNET, O. SANDRA and B. AUPERIN

Index 213

Contributors

ALAM, M.S.
Department of Biology, University of Southampton, Bassett Crescent East, Southampton SO9 3TU, UK
AUPERIN, B.
INRA, Laboratoire de Physiologie des Poissons, Campus de Beaulieu, Rennes 35042, France
CHENG, C.-H.C.
Department of Physiology, University of Illinois, 524 Burrill Hall, 407 S. Goodwin, Urbana 61801, USA
COSSINS, A. R.
Department of Environmental and Evolutionary Biology, University of Liverpool, PO Box 147, Liverpool L69 3BX, UK
CRAMB, G.
Molecular Endocrinology Group, Bute Medical Buildings, University of St Andrews, St Andrews, Fife KY16 9TS, UK
CUTLER, C.P.
School of Biological and Medical Sciences, University of St Andrews, Bute Medical Buildings, St Andrews, Fife KY16 9TS, UK
EL HAJ, A.J.
Department of Biological Sciences, University of Birmingham, PO Box 363, Birmingham B15 2TT, UK
ENNION, S.
Department of Anatomy and Developmental Biology, Royal Free Hospital School of Medicine, Rowland Hill Street, London NW3 2PF, UK
FAUCONNEAU, B.
INRA, Laboratoire de Physiologie des Poissons, Campus de Beaulieu, Rennes 35042, France
GAUVRY, L.
INRA, Laboratoire de Physiologie des Poissons, Campus de Beaulieu, Rennes 35042, France
GOLDSPINK, G.
Department of Anatomy and Developmental Biology, Royal Free

Hospital School of Medicine, Rowland Hill Street, London NW3 2PF, UK
GRACEY, A.Y.
Environmental Physiology Research Group, University of Liverpool, PO Box 147, Liverpool L69 3BX, UK
GRIFFIN, K.
Developmental Biology Research Centre, Randall Institute, King's College, 26–29 Drury Lane, London WC2B 5RL, UK
HAZON, N.
School of Biological and Medical Sciences, University of St Andrews, Gatty Marine Laboratory, St Andrews, Fife KY16 8LB, UK
HOLDER, N.
Developmental Biology Research Centre, Randall Institute, King's College, 20–29 Drury Lane, London WC2B 5RL, UK
IYENGAR, A.
Department of Biology, University of Southampton, Bassett Crescent East, Southampton SO9 3TU, UK
KOBOLÁK, J.
Institute for Molecular Genetics, Agricultural Biotechnology Center, Gödöllõ, Hungary
Present address: Department of Biology, University of Southampton, Bassett Crescent East, Southampton SO9 3TU, UK
LUKE, G.
Molecular Endocinology Group, Bute Medical Buildings, University of St Andrews, St Andrews, Fife KY16 9TS, UK
MACLEAN, N.
Department of Biology, University of Southampton, Bassett Crescent East, Southampton SO9 3TU, UK
MÜLLER, F.
Institute for Molecular Genetics, Agricultural Biotechnology Center, Gödöllõ, Hungary
Present address: Department of Biology, University of Southampton, Bassett Crescent East, Southampton SO9 3TU, UK
ORBAN, L.
Department of Biology, University of Southampton, Bassett Crescent East, Southampton SO9 3TU, UK
PATIENT, R.
Developmental Biology Research Centre, Randall Institute, King's College, 20–29 Drury Lane, London WC2B 5RL, UK
PEREZ, C.
INRA, Laboratoire de Physiologie des Poissons, Campus de Beaulieu, Rennes 35042, France

POPPLEWELL, A.
AFRC Centre for Genome Research, The University of Edinburgh, King's Buildings, West Mains Road, Edinburgh EH9 3JQ, UK

PRUNET, P.
INRA, Laboratoire de Physiologie des Poissons, Campus de Beaulieu, Rennes 35042, France

SANDERS, I.L.
Molecular Endocrinology Group, Bute Medical Building, University of St Andrews, St Andrews, Fife KY16 9TS, UK

SANDRA, O.
INRA, Laboratoire de Physiologie des Poissons, Campus de Beaulieu, Rennes 35042, France

TIKU, P.E.
Department of Environmental and Evolutionary Biology, University of Liverpool, PO Box 147, Liverpool L69 3BX, UK

WILLIAMS, D.W.
Department of Biology, University of Southampton, Bassett Crescent East, Southampton SO9 3TU, UK

XU, Q.
Developmental Biology Research Centre, Randall Institute, King's College, 26–29 Drury Lane, London WC2B 5RL, UK

Preface

Molecular biology is having a major impact in all areas of biological research. However, rather than being regarded as a separate discipline it should be seen as an extension to our existing repertoire of techniques, enabling us to study biological mechanisms at the whole organism, tissue, cellular and now at the molecular level. This volume presents some of the recent advances which have been made in understanding fundamental biological mechanisms in aquatic organisms by the application of molecular biology techniques.

In spite of homeostatic mechanisms, cellular environmental changes tend to be more extreme in aquatic animals as they cannot avoid external environmental changes such as temperature or salinity by simple behavioural means. The first four and the final chapter of this volume describe some of the recent advances made in understanding how adaptation to the environment is achieved at the level of gene expression. The first chapter (Cheng) describes how Antarctic notothenioid species of fish have adapted to their extremely cold environment by the production of antifreeze glycopeptides and provides new insights at the genomic level as to how heterogeneity of the antifreeze glycopeptides has arisen. In addition to long-term adaptation, some species of fish need to adapt to more acute seasonal changes in temperature. The chapter by Tiku *et al.* describes how the lipid composition of the cell membrane in carp is controlled by desaturases in response to cold temperature exposure, demonstrating a fascinating situation where transcription of a gene is increased by a decrease in the environmental temperature. The function of skeletal muscle is also acutely affected by changes in environmental temperature, and the chapter by Goldspink describes how changes in the expression of different isoforms of the myosin heavy chain genes in Carp are involved in maintaining locomotor function over a wide temperature range. As well having to overcome changes in environmental temperature, some species of fish have to contend with considerable changes in salinity, and the chapter by Cutler *et al.* describes how ion transporting proteins are involved in this adaptation. Prolactin has also been shown to be important in osmoregulation, and the chapter by Prunet *et al.* describes the molecular characterization of the receptor for this polypeptide hormone.

The next four chapters of the volume cover aspects of gene expression in the growth and development of aquatic organisms. El Haj describes some of the genes which are involved in the growth of crustaceans which represent an interesting group of animals in that they grow intermittently. The chapter by Xu *et al.* describes why the zebrafish has emerged as a key model for developmental biologists, showing that as well as being interesting subjects of study in their own right, aquatic organisms can provide useful experimental models for the study of vertebrate development as a whole. Most of the body mass of fish is composed of skeletal muscle and the major contractile protein in this tissue is the myosin heavy chain. The chapters by Ennion and Gauvry *et al.* describe the advances made in understanding the genes which code for the different isoforms of this protein and their differential expression in growth and development in carp and trout, respectively.

The transfer of engineered or foreign genes to produce transgenic animals offers, in many ways, a better system of studying gene expression than transfecting cells in culture, as the latter tend to behave as embryonic cells and frequently give erroneous results. Using transgenic animals, it should be possible to define the regulatory elements that are involved in tissue-specific, developmental stage-specific expression as well as the response elements for gene switching induced by the hormonal and physical signals associated with altered environmental conditions. In many respects the production of transgenic fish has been less successful than mammalian transgenics and this has mainly been due to the comparative lack of fish-derived gene regulatory sequences. A lot of the early work in transgeneic fish was performed with mammalian-derived gene regulatory sequences and, perhaps not surprisingly, the efficiency of these sequences in fish species was relatively poor. Recently there has been a rapid increase in the amount of data available for fish gene regulatory sequences and the chapters by Maclean and Müller deal with the techniques for assessing the efficiency of gene regulatory sequences in transgenic fish.

These are exciting times and, by using the available and emerging molecular biology methods, we can understand the basic mechanisms of growth, developmental and adaptation at a fundamental level.

S. ENNION
G. GOLDSPINK

C.-H.C. CHENG

Genomic basis for antifreeze glycopeptide heterogeneity and abundance in Antarctic fishes

Introduction

The evolution and expression of protein antifreezes in cold-water marine fishes which permitted them to thrive in otherwise lethally frigid marine habitats represents one of the most clear-cut and remarkable forms of cold adaptation that nature has invented. Antifreeze proteins (AFs) in polar fishes presumably evolved under the selective pressure of cold temperature as the polar oceans cooled to the freezing point of seawater (−1.9 °C) over their respective paleogeographic time scales. The Antarctic coastal waters today are perennially at −1.9 °C and ice-laden (Littlepage, 1965), while the Arctic and some north temperate waters experience similar conditions in boreal winters. Marine teleosts face the danger of freezing in these environments because they are hyposmotic to seawater; the salt content in their blood depresses the colligative or equilibrium freezing point only to about −0.7 °C (Prosser, 1973). They are thus supercooled with respect to ambient freezing seawater, and cannot avoid freezing in the presence of ice. Fishes living in cold waters generally have higher blood salt content; Antarctic fishes, for example, have enough salt to depress the colligative freezing point to −1.1 °C to −1.3 °C (DeVries, 1982; Ahlgren *et al.*, 1988) but this is still insufficient to prevent freezing. Presence of AFs in the blood and body fluids of AF-bearing fishes depress the freezing point further to a few tenths below −1.9 °C, preserving the body fluids in the liquid state (DeVries, 1982). Freezing point depression by AFs is via a non-colligative mechanism; AF molecules absorb to specific faces of ice crystals (Knight, Cheng & DeVries, 1991; Knight, Driggers & DeVries, 1993; Knight & DeVries, 1994) and inhibit ice growth through the Kelvin effect (Raymond & DeVries, 1977). Ice crystals become attached to fishes swimming in the ice-laden seawater or enter the intestinal fluid because they drink seawater for osmoregulation (DeVries & Cheng, 1992). Antifreeze proteins adsorb to these crystals and prevent them from growing at the temperatures the fish live in and thus prevent organismal freezing.

Aside from their unique function, antifreeze proteins as a single functional class of proteins are typified by two distinctive characteristics, structural diversity and molecular heterogeneity, and their high level of abundance in the fish especially in Antarctic fishes which experience persistent cold temperature extremes. The complexity of the molecular heterogeneity and the great abundance of antifreeze protein are illustrated in this chapter for the Antarctic notothenioid fishes, and the interesting gene structures and organization that give rise to these two characteristics are discussed.

Structural diversity and molecular heterogeneity of antifreeze proteins

Very different AFs in terms of protein sequence and secondary and tertiary structures have evolved in different fish taxa (Fig. 1). The Antarctic notothenioid fishes and several Arctic cods synthesize an alanine-rich glycopeptide antifreeze (DeVries, 1971; Van Voorhies, Raymond & DeVries, 1978; Fletcher, Hew & Joshi, 1982*a*; O'Grady, Ellory & DeVries, 1982a) which has a polyproline helix structure (Raymond, Radding & DeVries, 1977; Bush & Feeney, 1986). Other cold-water fishes synthesize one of three distinct types of peptide antifreeze. Type I AFPs (antifreeze peptide) are small alanine-rich, α-helical peptides of 3500–4500 D made by flounders, sculpins, and plaice (DeVries & Lin, 1977*a*; Hew *et al.*, 1985; Scott *et al.*, 1987; Chakrabartty *et al.*, 1988; DeVries & Cheng, 1992). Type II AFPs are cysteine-rich and β-structured peptides of about 14 000 to 17 000 D and are made by sea raven, smelt and herring (Ng & Hew, 1992;

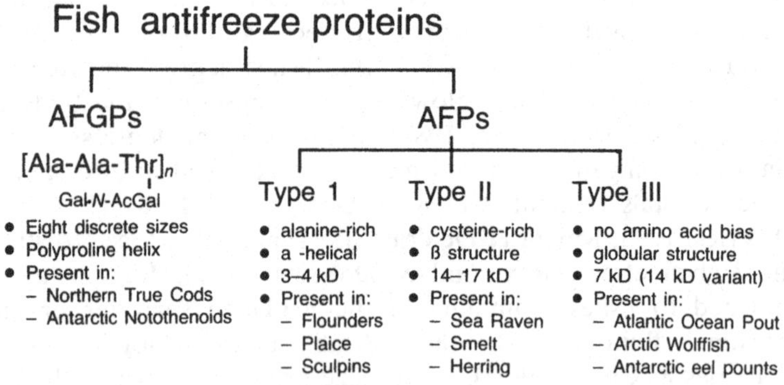

Fig. 1. Types of antifreeze proteins made by various polar and cold-water fishes.

Ewart & Fletcher, 1993). Type III AFPs are small peptides of about 6500 to 7000 D made by polar eel pouts, Atlantic ocean pout and wolffish (Schrag *et al.*, 1987; Hew *et al.*, 1988; Scott *et al.*, 1988; Cheng & DeVries, 1989) except for a 14 000 D two-AFP-domain variant made by an Antarctic eel pout (Wang, DeVries & Cheng, 1995*a*). Type III AFPs have no biased amino acid and are globular in overall structure (Sonnichesen *et al.*, 1993). These structurally very disparate AF molecules all interact with the same substrate, that is, ice, and perform the same function, which is paradoxical in view of the specificity observed of protein–substrate interactions in general.

Besides structural diversity, each given type of AF is synthesized not as a single molecular species but as a group of similar but distinct variants. In the case of AFPs, the number of variants within a given fish range from 2 to as many as 13. They are compositional variants differing from each other by several amino acids but are similar in size. An exception is the case of the Antarctic eel pout whose type III AFP compositional variants are made in two different sizes, 7000 D and 14 000 D (Wang *et al.*, 1995*a*). By contrast, the glycopeptide antifreezes occur as a family of polymers of varying length or size variants, but are all composed of essentially the same monomeric unit. This will be described in more detail below for the AFGPs (antifreeze glycopeptides) from the Antarctic notothenioid fishes.

Antifreeze protein abundance in fish

The AF-bearing fishes of the Arctic and north temperate oceans where summer water temperature can reach as high as 15 °C show seasonal variations in their circulatory levels of AFs, with low levels in the summer and high levels (10–15 mg/ml) in the winter (Petzel, Reisman & DeVries, 1980; Fletcher, Slaughter & Hew, 1982*b*; Fletcher *et al.*, 1984). By contrast, Antarctic fishes which experience perpetual freezing conditions express antifreeze proteins constitutively, and maintain them at constant high circulatory concentrations. The AFGP levels in the blood and body fluids of Antarctic notothenioid fishes are 35–40 mg/ml, the highest of all AF-bearing fishes (DeVries & Lin, 1977*b*; DeVries, 1982). This level of abundance ranks with that of the most abundant serum protein, albumin, but since AFs in general are much smaller molecules than serum albumin, their abundance on a molar basis is in fact much greater. The high level of production and molecular heterogeneity of AFs arise in part from the fact that they are encoded by large multigene families, which have been characterized for some of the northern AFP-bearing fishes (Scott, Hew & Davies, 1985; Hew *et al.*, 1988; Scott *et al.*, 1988), and recently for the Antarctic eel pout (Wang, DeVries & Cheng, 1995*b*). The molecular heterogeneity of the AFGPs has been mostly characterized in the Antarctic notothenioids, and the genomic bases for the heterogeneity and the tremendous output of AFGPs in these fishes are recently being studied in our laboratory and are beginning to become clear.

AFGPs (antifreeze glycopeptides) of Antarctic cods

Antarctic notothenioid fishes (Suborder Notothenioidei, Order Perciform) are commonly known as Antarctic cods but they are neither cods nor are they related to northern cods (Order Gadiform). Antarctic cods comprise the majority of fish fauna in the Antarctic Ocean particularly in the shelf regions (Eastman, 1991, 1993). Antifreeze protein was first discovered in Antarctic cods (DeVries, 1968, 1971), and was found to be a series of glycosylated proteins. Similar antifreeze glycoproteins were subsequently found in the northern cods (Van Voorhies *et al.*, 1978; Fletcher *et al.*, 1982*a*; O'Grady *et al.*, 1982*b*).

AFGP size and compositional heterogeneity

The first AFGPs were isolated from the small cryopelagic Antarctic cod, *Perciform borchgrevinki* by DEAE–cellulose anion-exchange chromatography of the fish plasma. They were shown to be a series of polymers composed of varying repeats (n) of the glycotripeptide monomer (Ala–Ala–Thr–)$_n$ with the Thr residue *O*-linked to the disaccharide, galactosyl-*N*-acetogalactosamine (DeVries, 1971; Shier, Lin & DeVries, 1972, 1975), and ending with the dipeptide–Ala(Pro)–Ala (Fig. 2). Early characterizations uncovered eight distinct sizes of AFGPs (DeVries, Komatsu & Feeney, 1970; Komatsu, DeVries & Fenney, 1970) the largest one was named AFGP1, and the smallest, AFGP8, with a MW of about 34 000 D (n = 56 repeats) and 2600 D (n = 4 repeats), respectively (Figs. 2 and 3(*a*))(DeVries, 1971, 1982). By fluorescent labelling and electrophoresis of AFGPs on gradient polyacrylamide gel, many more sizes of AFGPs were detected. Fig. 3(*b*) shows the many AFGP size variants from the blood of individual specimens of *P. borchgrevinki*, and of the giant Antarctic cod *Dissostichus mawsoni* resolved on gradient gel. Comparing Figs. 3(*a*) and (*b*), AFGP 7 and 8 remain as single bands, or a single size, but AFGP6 which was originally detected as a single band resolved into six bands or six sizes. Similarly, the single bands of AFGP5, 4 and 3 all resolve into groups of several bands. The number of bands in each lane totals at least 18, thus there are at least 18 different AFGP size variants in each fish based

NH_2–[Ala(Pro) – Ala – Thr–]$_n$ Ala(Pro)–Ala–COOH
/
***N*-Acetylgalactosamine**
/
Galactose

AFGP	MW (Daltons)	Number of glycotripeptide repeats (*n*)
1	33 700	56
2	28 800	47
3	21 500	35
4	17 000	27
5	10 500	17
6	7 900	12
7	3 500	5
8	2 600	4

Fig. 2. Structure of AFGPs (antifreeze glycopeptides) 1 to 8 from Antarctic notothenioid fishes and their molecular mass determined by sedimentation equilibrium centrifugation. AFGPs are composed of varying numbers (*n*) of the glycotripeptide monomer, Ala–Ala–Thr– with the disaccharide *O*-linked to each Thr. Pro for Ala substitutions are present in the small AFGPs 6 to 8.

on gel mobilities alone. Fig. 3(*b*) also shows that there is intraspecific as well as interspecific differences in the size heterogeneity of the large AFGPs, 1 to 5, indicated by the different banding patterns of these sizes among the individuals within each fish species and between the two species. In addition, the largest AFGP of *D. mawsoni* is much larger than that (AFGP1) of *P. borchgrevinki* as indicated by its much slower mobility in the gel. The first and last lane in Fig. 3(*b*) represents physiological abundance of the AFGP variants in each fish; the smallest AFGPs, 7 and 8, comprise over 70% of the circulating AFGPs, and AFGPs1 to 6 comprise the remainder.

In addition to size heterogeneity, the small AFGPs, 6 to 8, differ slightly from the large AFGPs 1 to 5 in amino acid composition, in that some of the Alas are replaced by Pros (Lin, Duman & DeVries, 1972; Morris *et al.*, 1978). The Pro for Ala replacements always occur at the first Ala following a Thr residue. The single size of AFGP8 from the small cod, *P. borchgrevinki*, was found to consist of three

(*a*)

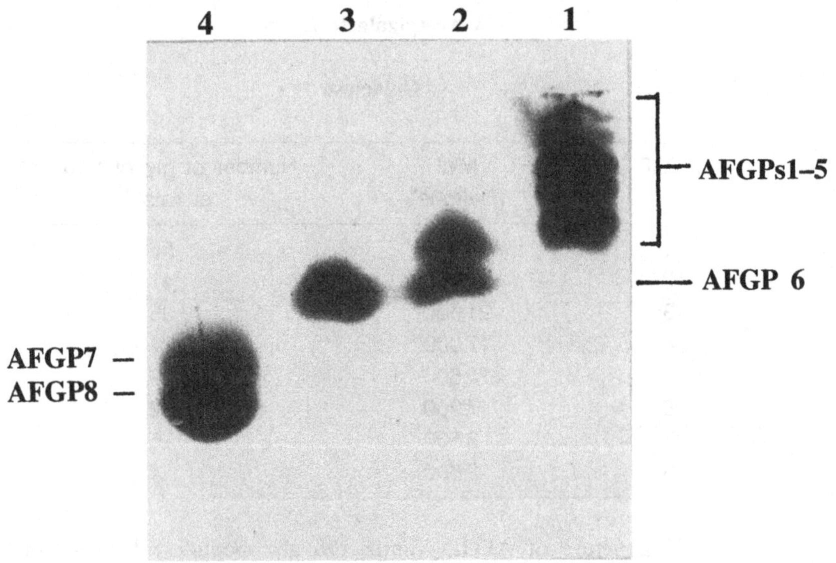
4
3
2
1
AFGPs1–5
AFGP 6
AFGP7
AFGP8

(*b*)

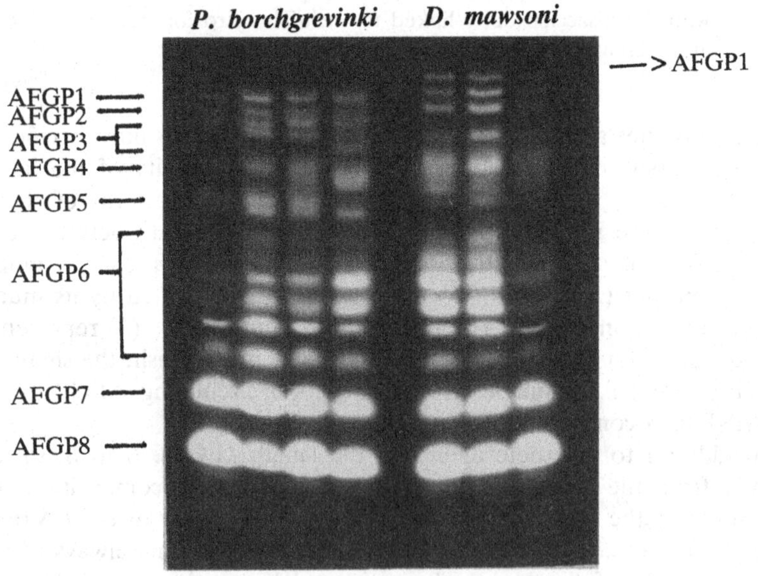
P. borchgrevinki
D. mawsoni
>AFGP1
AFGP1
AFGP2
AFGP3
AFGP4
AFGP5
AFGP6
AFGP7
AFGP8

7 13
Ala-Ala-Thr-Ala-Ala-Thr-**Pro**-Ala-Thr-Ala-Ala-Thr-**Pro**-Ala

10 13
Ala-Ala-Thr-Ala-Ala-Thr-Ala-Ala-Thr-**Pro**-Ala-Thr-**Pro**-Ala

13
Ala-Ala-Thr-Ala-Ala-Thr-Ala-Ala-Thr-Ala-Ala-Thr-**Pro**-Ala

Fig. 4. Amino acid sequence of the three compositional variants of AFGP8 with the positions of Pro for Ala substitutions indicated (from Morris *et al.*, 1978).

compositional variants (Fig. 4), one with a single Pro for Ala replacement in position 13, and the other two with two replacements in positions 7 and 10, and 10 and 13, respectively (Morris *et al.*, 1978). Assuming AFGP7 and the six sizes of AFGPs6 also have three Pro variants each, the total number of size and compositional variants in the fish reaches at least 34. This level of complexity is the highest among all characterized types of antifreeze proteins.

Why so many sizes of AFGPs?

The occurrence of a large number AFGP variants in the Antarctic cods raises two questions, why are these different sizes needed and how are all the size and compositional variants encoded and synthesized? The answer to the first question is not entirely clear. It is known, however, that there is a direct relationship between molecular size and antifreeze activity; at the same concentration, the large AFGPs 1 to 5 produce a larger freezing point depression than the small AFGPs 7 and 8 (Schrag, O'Grady & DeVries, 1982). And, using large single

---

Fig. 3. Resolution of AFGPs on polyacrylamide gel. (*a*) Initial method: eight sizes of AFGPs from *P. borchgrevinki* were identified on single percentage gel treated with 5% α-naphthol in 90% ethanol and concentrated sulfuric acid which stains the disaccharides. Lanes 1 to 4 represent AFGP fractions in the order of elution from DEAE anion-exchange chromatography of fish plasma. (*b*) Current method: at least 18 bands (sizes) are identified on gradient electrophoresis of AFGPs labelled with a fluorescent tag (fluorescamine). AFGP3, 4, 5 and 6 resolved into multiple bands. The bands corresponding to initial nomenclature of AFGP sizes in (*a*) are as indicated. Each lane represents AFGPs from an individual fish. Lanes 1 to 4: *P. borchgrevinki*; lanes 5 to 7: *D. mawsoni*. Lane 1 and lane 7 show the physiological abundance of the AFGP variants in the respective fish.

ice crystals, it was shown that large AFGPs are much more effective inhibitors of the ice crystal growth than the small AFGPs (Raymond, Wilson & DeVries, 1989). However, not every fluid compartment in the fish receives the full complement of AFGP sizes. The intestinal fluid of Antarctic cods contains only the small AFGPs 7 and 8, where they reach very high concentrations (40 mg/ml) to prevent freezing initiation by ice crystals inadvertently ingested by the fish along with drinking ice-laden seawater (O'Grady *et al.*, 1982*a*). AFGPs 7 and 8 enter the intestine tract presumably by translocation from the liver to the gall bladder which empties into the intestine (Cheng & DeVries, unpublished results), and their small size may facilitate the transport process. Thus there appears to be at least a functional selection rationale for the synthesis of both the large and small size variants.

AFGP polyprotein genes

How the complex AFGP heterogeneity arises resides in the study and analyses of the genetic blueprints for these molecules. *A priori* it is intuitively certain that the large number of AFGP variants are encoded by a family of many genes. The caveat is, if each gene encodes a single peptide precursor of each AFGP variant, the gene family would be one of many very small genes, since the majority (≥70%) of the circulating antifreeze in the Antarctic cods are the small AFGPs 7 and 8 with only 17 and 14 residues, respectively, in their peptide backbone, corresponding to only 51 and 42 nucleotides, respectively, of coding sequence. It is not obvious how very small peptides could be efficiently expressed to reach the high levels of mature AFGPs observed in the blood. Characterization of the first AFGP gene isolated from a genomic library of the Antarctic cod, *Notothenia coriiceps* (Hsiao *et al.*, 1990) provided some answers; the AFGP gene encodes not just a single but many AFGP molecules.

The *N. coriiceps* gene (abbrev. *NC-AFGP*) is almost 2500 nucleotides in length in the coding sequence and contains no introns. It encodes a putative signal peptide of 37 residues in length, and a large precursor antifreeze protein. The precursor protein contains 46 AFGP molecules successively linked in direct tandem by 3-residue spacers with the highly conserved sequence of Leu(Phe)–Xaa–Phe (Figs. 5, 6). Of the 46 AFGP molecules, 43 are AFGP8 variants, 2 are AFGP7 variants, and the remaining 1 has 3 tripeptide repeats and hence 1 repeat smaller than AFGP8. There are 5 different compositional variants in the 43 molecules of AFGP8; the two predominant ones are a single Pro for Ala substitution at position 10 (20 copies), and 2 substitutions at

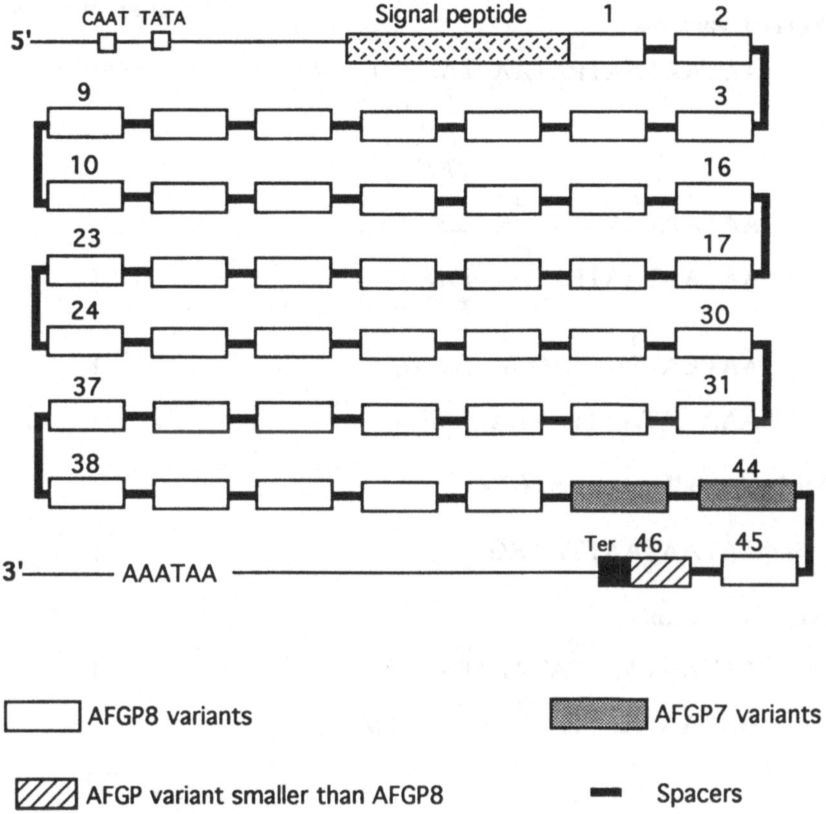

Fig. 5. The organization of the AFGP polyprotein gene of *N. coriiceps* (*NC–AFGP* gene). Putative CAAT and TATA boxes, signal peptide, termination codon and polyadenylation signal (AATAAA) are as indicated. The 46 tandem copies of AFGP coding sequences represented by the series of rectangular boxes linked by 9-bp spacers (thick line) are numbered. Copies number 1 to 42, and 45 encode AFGP8 variants, 43 and 44 encode AFGP7 variants, and copy 46 at the 3' end encodes a variant with 3 tripeptide repeats, 1 repeat smaller than AFGP8.

positions 7 and 13 (17 copies). There are 4 copies with 2 substitutions, at positions 10 and 13, and a single copy with 3 substitutions, at positions 7, 10 and 13. The remaining AFGP8 copy has a single substitution at position 10 and has an Ala instead of a Thr in the first tripeptide repeat. The last copy in the precursor protein is the variant

			Number of copies
AFGP8 variants			
10			
AATAATAATPATAA	*LNF (15)*		20
	LHF (1)		
	FNF (2)		
	CNF (2)		
7 13			
AATAATPATAATPA	*LIF (17)*		17
10 13			
AATAATAATPATPA	*FHF (3)*		4
	FNF (1)		
7 10 13			
AATAATPATPATPA	*LIF (1)*		1
10			
AAAAATAATPATAA	*LNF (1)*		1
Variant smaller than AFGP8			
AATAATAATTAARG			1
AFGP7 variants			
10 16			
AATAATAATPATAATPA		*LIF (1)*	1
10 14 16			
AATAATAATPATPATPA		*LIF (1)*	1
		Total	**46**

Fig. 6. Amino acid sequence and number of copies of the AFGP variants encoded in the *NC–AFGP* polyprotein gene. The 3-residue spacer following each variant and its frequency of occurrence are in italics. The number above Pro residues indicates their positions in the mature AFGPs.

that is smaller than AFGP8 and it has an additional dipeptide, –Arg–Gly, at its C-terminus. Each of the two AFGP7 is a different variant, with 2 (positions 10 and 16) and 3 (positions 10, 13 and 16) Pro for Ala substitutions, respectively (Fig. 6).

Several AFGP genes from the giant Antarctic cod *D. mawsoni* have recently been characterized in our laboratory. The giant cod AFGP genes have the same polyprotein structure as that of the *NC–AFGP* gene. Most of the genes characterized contain multiple copies of coding sequence for distinct variants of the smallest AFGP8, several copies for 7, and one or two copies for AFGP6. The last copy in each gene ends with the conserved dipeptide, –Arg–Gly. One gene has been identified to encode, among several copies of the small AFGPs 6–8, two large AFGPs one of which is larger than AFGP1 of *P. borchgrevinki*, and the other in the range of AFGP4 (manuscript in preparation). The high frequency of encountering genes that predominantly encode the small AFGPs 8 and 7 in the isolation of AFGP genomic clones from the library of both *N. coriiceps* and *D. mawsoni* is consistent with the fact that AFGPs 8 and 7 are the most abundant variants in the blood (Fig. 3(*b*)). The gene sequences of these two fishes also show that each and every size and compositional AFGP variant is distinctly encoded, not as single genes, but as distinct copies within large polyprotein genes.

Maturation of AFGP polyproteins

The highly conserved 3-residue spacers that link the multiple AFGP molecules in the large AFGP polyprotein have the sequence of Leu(Phe)–Xaa–Phe. Of the 45 spacers in the *NC*–AFGP gene, 20 are Leu–Ile–Phe (LIF), 16 Leu–Asn–Phe (LNF), 3 Phe–Asn–Phe (FNF), 3 Phe–His–Phe (FHF), 2 Cys–Asn–Phe (CNF), and 1 Leu–His–Phe (LHF) (Fig. 6). Thus, the first residue of all spacers except two is Leu or Phe, the third residue is always Phe, and the middle one is predominantly Ile or Asn. None of the spacer residues have been identified in mature AFGPs. The protease chymotrypsin is known to cleave preferentially on the carboxyl side of Leu, Phe, Trp and Tyr, and to a less extent that of Asn, His, Met and Gln (Croft, 1980). In Northern analysis of AFGP mRNAs from *N. coriiceps*, they are found to be large messages of 1 to several kb in size, and one of them, about 3 kb, matches the predicted mRNA size of the *NC–AFGP* gene (Hsiao *et al.*, 1990). These data taken together suggest that the *NC–AFGP* gene (applicable to other *AFGP* genes as well) is transcribed into a large mRNA and translated into a large polyprotein, which is then cleaved at the spacer sequences post-translationally by chymotrypsin-like protease to give 46 AFGPs; presumably a carboxypeptidase-like protease removes the remaining single or two residues to produce the mature AFGPs.

It is not known at what point in the AFGP synthesis when cleavage of the polyprotein occurs. Since each nascent prepro-AFGP chain has a single signal peptide, it must enter as a single entity into RER (rough endoplasmic reticulum) where the signal peptide is removed by signal peptidase, followed by transfer of the proAFGP in vesicles to the *cis* membraneous sacs of the Golgi complex. Glycosylation of secretory proteins takes place in the Golgi complex. Small molecular weight peptides in general are known to glycosylate poorly (Hill *et al.*, 1977), and thus it is logical to assume that glycosylation of AFGP takes place at the stage of the large polyprotein. Golgi is also known to contain proteolytic enzymes that participate in maturation of proproteins. Proinsulin for example is sorted into secretory vesicles that bud off from the trans pole of the Golgi complex, and cleaved in the vesicles to remove the C chain by trypsin-like protease to produce active insulin (Orci *et al.*, 1987). Thus cleavage of the glycosylated AFGP polyprotein may also take place in the Golgi complex or vesicles, although this remains to be demonstrated.

Multigene family and multiple *AFGP* per gene organization creates a large gene dosage

Southern blot of genomic DNA of 3 Antarctic cods, *N. coriiceps, D. mawsoni* and the haemoglobinless *Chaenocephalus aceratus* (family Channichthyidae) hybridized to a probe made with the *NC–AFGP* gene shows many positive bands indicating that AFGPs are encoded by a sizeable gene family across the notothenioids (Fig. 7). The 4-bp restriction enzyme *TaqI* does not cut into the repetitive coding sequence of *AFGP* genes, but cuts right outside of these genes where the sequence is random and the probability of encountering its recognition sequence is high. Thus the hybridizing bands of the *TaqI* digest lanes represent complete *AFGP* genes. Some of the *TaqI* positive bands have strong hybridization intensity, indicating the co-migration of a number of positive fragments of the same size. The polyprotein nature of the *AFGP* genes means each gene is equivalent to many genes. The single *NC–AFGP* gene alone is equivalent to 46 single-peptide genes. This multigene family, and multiple copies of *AFGP*s per gene organization can create a very large effective gene dosage, which is consistent with the high circulatory abundance of the mature AFGPs in the Antarctic cods. Besides augmenting the gene dosage, the unique *AFGP* gene structures may contribute to reduction in the energetic costs of AFGP biosynthesis in two ways. The fact that *AFGP* genes are intronless eliminates the need for mRNA splicing and thus enhances transcriptional efficiency. In the secretory pathway, the transport of a single AFGP polyprotein into the RER and Golgi complex is equivalent to having transported many individual AFGP precursors. The energetic saving may be significant in the low temperature environment the Antarctic cods live in.

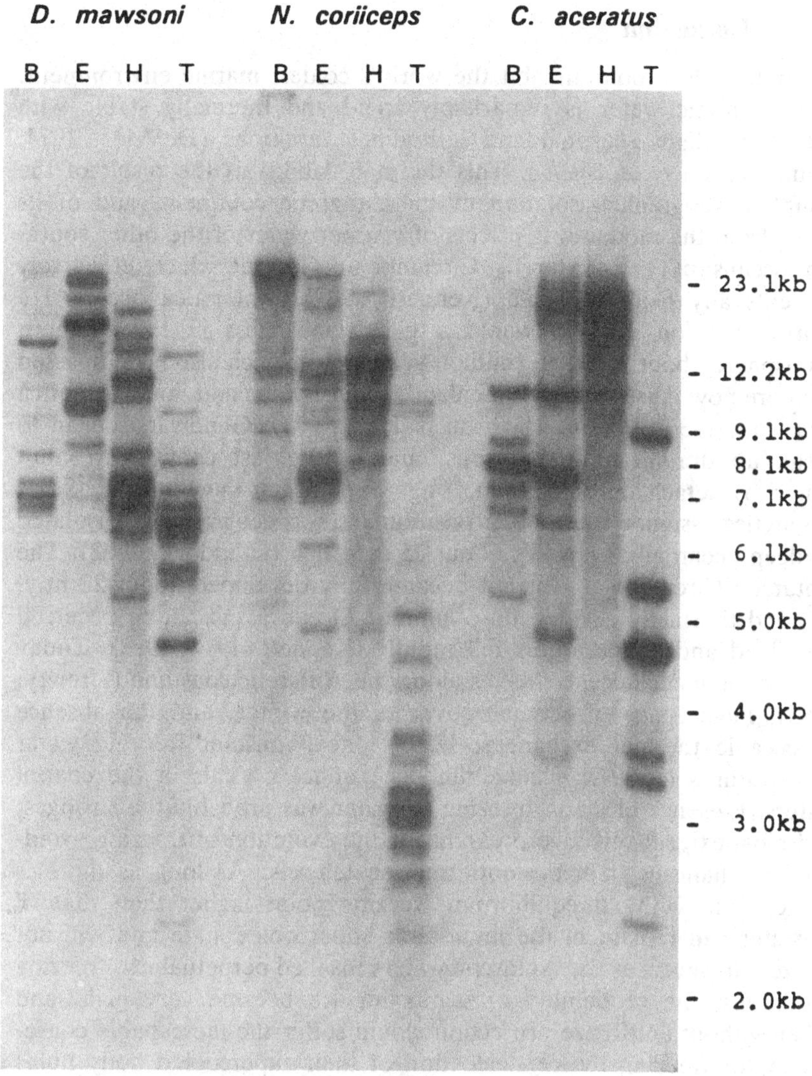

Fig. 7. Genomic Southern blot of three Antarctic notothenioid fishes showing large AFGP multigene families. Blot was hybridized to a probe made with the *NC–AFGP* polyprotein gene. Each lane contains 100 μg of restricted genomic DNA. B: BamHI, E: EcoRI, H: HindIII, T: TaqI.

Discussion

Antarctic fishes today inhabit the world's coldest marine environment. The Antarctic water is remarkably frigid and thermally stable with only very slight seasonal and latitudinal variations (DeVries, 1974; Holm-Hansen *et al.*, 1977). This thermal stability is the result of the complete geographic isolation of the Antarctic continent, and of its water from the moderating effects of warmer water of the other southern oceans by the Antarctic Circumpolar Current which effectively prevents any heat exchange (Kennett, 1982). Antarctica occupied a central position of Gondwana, a giant land mass in the southern hemisphere about 200 m.y. (million years) ago, which also encompassed what are now Australia, New Zealand, India, Africa and South America (Du Toit, 1937; Barron, Harrison & Hay, 1978). Gondwana began to break up during mid-Cretaceous, and as the last continent (South America) detached about 38 m.y. ago opening up the Drake Passage, Antarctica assumed south polar position and was geographically isolated by deep ocean all around by about 25 m.y. ago (Craddock, 1982). The Antarctic Circumpolar Current became fully developed about 22 m.y. ago and thermally isolated the Antarctic Ocean as glaciation occurred over land and temperatures fell rapidly (Kennett, 1977, 1982). Today the predominance of ice shelves along the Antarctic coastline (Drewry, 1983), huge spans of sea ice cover in the winter, and the absence of oceanic thermal exchanges, lead to near-constant freezing water temperatures (−1.9 °C) and abundance of ice crystals in the coastal waters. Presence of ice in freezing seawater was probably the strongest and most exigent selective pressure for the evolution of freezing avoidance mechanisms in hyposmotic marine teleosts. As long as there is no ice, fish with an equilibrium freezing point higher than that of seawater can remain in the metastable supercooled state and will not freeze. However, as the Antarctic waters reached perpetual icy, freezing conditions, the probability of contacting ice became very high, and fishes without antifreeze protection would suffer the inescapable consequence of freezing by ice nucleation of their supercooled body fluids followed by death. The evolution of antifreeze proteins in the Antarctic marine teleosts enabled them to successfully colonize their ice-laden marine environment. Particularly successful are the fishes of the suborder Notothenioidae, which had radiated into a wide range of habitats, from benthic to different levels in the water column, and became the most diverse monophyletic group of fishes with a wide assortment of body sizes and forms as well as unique physiologies such as aglomerularism, haemoglobinless blood (Channicthyids), and neutral buoyancy (Eastman & DeVries, 1982, 1986*a*, *b*; Eastman, 1991, 1993). The present-day notothenioids comprise the predominant taxon in the Antarctic fish fauna, representing 95 of the 174 (54.5%) known species in the Antarctic continent shelf and shelf slope regions (Eastman, 1993). In parallel with their ecological success, the notothenioids appear to have evolved a particularly efficient system of expressing antifreeze proteins for freezing protection – a family of many intronless, AFGP polyprotein genes. Absence of intron enhances transcriptional efficiency, and the unique polyprotein gene structure – multiple copies of AFGP peptide coding sequences linked in direct tandem by small spacers that constitute cleavage sites, produces many AFGP molecules in a single round of transcription and translation. One would expect it to be extremely inefficient and energetically costly if the very small AFGPs, the 14-residue AFGP8 and the 17-residue AFGP7 variants, which make up the majority (≥70%) of circulatory AFGPs were to be endowed with a signal peptide, transcribed, translated and secreted individually. There are other examples of polyprotein genes in nature that generate multiple peptides of either identical sequence or related sequences, or peptides of different biological functions (for review see Douglass, Civelli & Herbert, 1984), but all have much fewer copies of encoded peptides (maximally 8) than the AFGP polyprotein genes.

The duplications of AFGP copies that give rise to the numerous iterations of AFGP coding sequences within each polyprotein gene, together with duplications of complete genes to increase the size of the gene family apparently must have been targeted towards increasing the gene dosage for production of AFGPs *en masse* to meet the environmental demand. Large gene families and thus large gene dosage are characteristic of gene products that are needed in abundance. A prime example of this is the histone proteins which occur in roughly the same mass as the DNA per cell. The size of histone gene families range from 10 to 20 genes per histone protein in birds and mammals to as many as 500 to 800 genes per histone in sea urchins and newt (Hentschel & Birnstiel, 1981). Antifreeze gene families of several AFP-bearing fishes have been characterized and they are sizeable. Type I *AFP* gene family of the winter flounder is estimated to contain about 40 genes (Scott, Hew & Davies, 1985), and type III *AFP* gene family of the Atlantic ocean pout (Hew *et al.*, 1988), the Atlantic wolffish (Scott *et al.*, 1988), and one of the Antarctic eel pout (Wang *et al.*, 1995*a*, *b*) contain about 180, 80–85 and 43 member genes, respectively. In all cases, however, each gene encodes a single AFP. Besides the two different kinds of duplications that increased gene dosage mentioned above, a third type of duplication presumably generated the

AFGP size heterogeneity. AFGPs are composed of a very simple short repeat, Ala–Ala–Thr, encoded by a 9-bp sequence. Once this simple sequence is duplicated, slippage pairing could easily occur, and through multiple rounds of slippage duplication, longer and longer lengths of AFGP could be generated for a given copy of coding sequence within a polyprotein gene. It appears then there is a simple answer to the question of why there are so many sizes of AFGPs. It may merely be the result of the conducive nature of the simple repetitive sequence of AFGP towards slippage replication, and the longer AFGPs just happen to have greater antifreeze activity and are retained in the genome. This process, in fact, may still be operative today as suggested by the intraspecific variability in the sizes of the large AFGPs seen on polyacrylamide gel electrophoresis (Fig. 3(*b*)).

In conclusion, the evolution of AFGPs in Antarctic notothenioids permitted them to successfully colonize their perpetual frigid marine environment and radiate into diverse ice-laden inhabitats to become the dominant fish taxa in the Antarctic Ocean. Their multigene family and multiple copy number per gene organization provides a large gene dosage for high level of AFGP production, as well as the basis for the observed AFGP size heterogeneity.

Acknowledgement

This work was supported in part by grant National Science Foundation Grant Office of Polar Programs 93-17629.

References

Ahlgren, J.A., Cheng, C.-H.C., Schrag, J.D. & DeVries, A.L. (1988). Freezing avoidance and the distribution of antifreeze glycopeptides in body fluids and tissues of Antarctic fish. *Journal of Experimental Biology*, **137**, 549–63.

Barron, E.T., Harrison, C.G.A. & Hay, W.W. (1978). A revised reconstruction of the southern continents. *Transactions of the American Geophysical Union*, **59**, 436–9.

Bush, C.A. & Feeney, R.E. (1986). Conformation of the glycotripeptide repeating unit of antifreeze glycoprotein of polar fish as determined from the fully assigned proton NMR spectrum. *Journal of Peptide and Protein Research*, **28**, 386–97.

Chakrabartty, A., Hew, C.L., Shears, M. & Fletcher, G. (1988). Primary structures of the alanine-rich antifreeze polypeptides from grubby sculpin (*Myoxocephalus aenaeus*). *Canadian Journal of Zoology*, **66**, 403–8.

Cheng, C.-H.C. & DeVries, A.L. (1989). Structures of antifreeze peptides from the antarctic fish *Austrolycicthys brachycephalus. Biochimica et Biophysica Acta*, **997**, 55–64.

Craddock, C. (1982). Antarctica and Gondwanaland. In *Antarctic Geoscience (Proceedings of the Symposium on Antarctic Geology and Geophysics)*, ed. C. Craddock, pp. 3–13. Madison: University of Wisconsin Press.

Croft, L.R. (1980). *Introduction to Protein Sequence Analysis*. Chichester: John Wiley & Sons.

DeVries, A.L. (1968). Freezing resistance in some Antarctic fishes. PhD Thesis. Stanford University, Stanford, California.

DeVries, A.L. (1971). Glycoproteins as biological antifreeze agents in Antarctic fishes. *Science*, **172**, 1152–5.

DeVries, A.L. (1974). Survival at freezing temperatures. In *Biochemical and Biophysical Perspectives in Marine Biology*, ed. D.C. Malins & J.R. Sargent, vol. 1, pp. 290–330. London: Academic Press.

DeVries, A.L. (1982). Biological antifreeze agents in coldwater fishes. *Comparative Biochemistry and Physiology*, **73A**, 627–40.

DeVries, A.L., Komatsu, S.K. & Feeney, R.E. (1970). Chemical and physical properties of freezing point-depressing glycoproteins from Antarctic fishes. *Journal of Biological Chemistry*, **245**, 2901–13.

DeVries, A.L. & Lin, Y. (1977*a*). Structure of a peptide antifreeze and mechanism of absorption to ice. *Biochimica et Biophysica Acta*, **495**, 388–92.

DeVries, A.L. & Lin, Y. (1977*b*). The role of glycoprotein antifreezes in the survival of Antarctic fishes. In *Adaptation within Antarctic Ecosystems (Proceedings of the Third Symposium on Antarctic Biology)*, ed. G.A. Llano, pp. 439–458. Houston: Gulf Publishing.

DeVries, A.L. & Cheng, C.-H.C. (1992). The role of antifreeze glycopeptides and peptides in the survival of cold water fishes. In *Water and Life: Comparative Analysis of Water Relations at the Organismic, Cellular, and Molecular Levels*, ed. G.N. Somero, C.B. Osmond & C.L. Bolis, pp. 301–315. Berlin, Heidelberg: Springer-Verlag.

Douglass, J., Civelli, O. & Herbert, E. (1984). Polyprotein gene expression: generation of diversity of neuroendocrine peptides. *Annual Review of Biochemistry*, **53**, 665–715.

Drewry, D.J. (1983). *Antarctica: Glaciological and Geophysical Folio*. Cambridge: Scott Polar Research Institute.

Du Toit, A.L. (1937). *Our Wandering Continents*. Edinburgh: Oliver and Boyd.

Eastman, J.T. (1991). Evolution and diversification of Antarctic notothenioid fishes. *American Zoology*, **31**, 93–109.

Eastman, J.T. (1993). *Antarctic Fish Biology*. San Diego: Academic Press.

Eastman, J.T. & DeVries, A.L. (1982). Buoyancy studies of notothenioid fishes in McMurdo Sound, Antarctic. *Copeia*, **1982(2)**, 385–93.

Eastman, J.T. & DeVries, A.L. (1986*a*). Renal glomerula evolution in Antarctic notothenioid fishes. *Journal of Fish Biology*, **29**, 649–62.

Eastman, J.T. & DeVries, A.L. (1986*b*). Antarctic fishes. *Scientific American*, **254**, 106–14.

Ewart, K.V. & Fletcher, G.L. (1993). Herring antifreeze protein: primary structure and evidence for a C-type lectin evolutionary origin. *Molecular Marine Biology and Biotechnology*, **2**, 20–27.

Fletcher, G.L., Hew, C.L. & Joshi, S.B. (1982*a*). Isolation and characterization of antifreeze glycoproteins from the frost fish *Microgadus tomcod*. *Canadian Journal of Zoology*, **60**, 348–55.

Fletcher, G.L., Slaughter, D. & Hew, C.L. (1982*b*). Seasonal changes in the plasma levels of glycoprotein antifreeze, Na^+, Cl^-, and glucose in Newfoundland Atlantic cod (*Gadus morhua*). *Canadian Journal of Zoology*, **60**, 1851–4.

Fletcher, G.L., Hew, C.L., Li, X.M., Haya, K. & Kao, M.H. (1984). Year-round presence of high levels of plasma antifreeze in a temperate fish, ocean pout (*Macrozoarces americanus*). *Canadian Journal of Zoology*, **63**, 488–93.

Hentschel, C.C. & Birnstiel, M.L. (1981). The organization and expression of histone gene families. *Cell*, **25**, 301–13.

Hew, C.L., Joshi, S., Wang, N.-C., Kao, M.-H. & Ananthanarayanan, V.S. (1985). Structures of shorthorn sculpin antifreeze polypeptides. *European Journal of Biochemistry*, **151**, 167–72.

Hew, C.L., Wang, N.-C., Joshi, S., Fletcher, G.L., Scott, G.K., Hayes, P.H. & Buettner, B. (1988). Multiple genes provide the basis for antifreeze protein diversity and dosage in the ocean pout, *Macrozoarces americanus*. *Journal of Biological Chemistry*, **263**, 12049–55.

Hill, H.D. Jr., Schwyzer, M., Steinman, H.N. & Hill, R.L. (1977). Ovine submaxillary mucin. Primary structure and peptide substrates of UDP-*N*-acetyl-galactosamine:mucin transferase. *Journal of Biological Chemistry*, **252**, 3799–804.

Holm-Hansen, O., El-Sayed, S.Z., Ranceschini, G.A. & Cuhel, R.L. (1977). Primary production and the factors controlling phytoplankton growth in southern ocean. In *Adaptations within Antarctic Ecosystems*, ed. G.A. Llano, pp. 11–50. Houston: Gulf.

Hsiao, K., Cheng, C.-H.C., Fernandes, I.E., Detrich, H.W. & DeVries, A.L. (1990). An antifreeze glycopeptide gene from the antarctic cod *Notothenia coriiceps neglacta* encodes a polyprotein of high peptide copy number. *Proceedings of the National Academy of Science*, **87**, 9265–9.

Kennett, J.P. (1977). Cenozoic evolution of Antarctic glaciation, the circum-Antarctic Ocean, and their impact on global paleoceanography. *Journal of Geophysical Research*, **82**, 3841–60.

Kennett, J.P. (1982). *Marine Geology*. New Jersey: Prentice-Hall.

Knight, C.A., Cheng, C.-H.C. & DeVries, A.L. (1991). Adsorption of α-helical antifreeze peptides on specific ice crystal planes. *Biophysical Journal*, **56**, 409–18.

Knight, C.A., Driggers, & DeVries, A.L. (1993). Adsorption of fish antifreeze glycopeptides to ice, and effects on ice crystal growth. *Biophysical Journal*, **64**, 252–9.

Knight, C.A. & DeVries, A.L. (1994). Effects of a polymeric, non-equilibrium "antifreeze" upon ice growth from water. *Journal of Crystal Growth*, **143**, 301–10.

Komatsu, S.K., DeVries, A.L. & Feeney, R.E. (1970). Studies of the structure of freezing point-depressing glycoproteins from an Antarctic Fish. *Journal of Biological Chemistry*, **245**, 2909–13.

Lin, Y., Duman, J.G. & DeVries, A.L. (1972). Studies on the structure and activity of low molecular weight glycoproteins from an Antarctic fish. *Biochemical and Biophysical Research Communication*, **46**, 87–92.

Littlepage, J.L. (1965). Oceanographic investigations in McMurdo Sound, Antarctic. In *Antarctic Research Series, Vol. 5. Biology of Antarctic Seas II*, ed. G.A. Llano, pp. 1–37. Washington: American Geophysical Union.

Morris, H.R., Thompson, M.R., Osuga, D.T., Ahmed, A.T., Chan, S.M., Vandenheede, J.R. & Feeney, R.F. (1978). Antifreeze glycoproteins from the blood of an Antarctic fish. *Journal of Biological Chemistry*, **253**, 5155–62.

Ng, N.F. & Hew, C.L. (1992). Structure of an antifreeze peptide from the sea raven. *Journal of Biological Chemistry*, **267**, 16069–75.

O'Grady, S.M., Ellory, J.C. & DeVries, A.L. (1982*a*). Protein and glycoprotein antifreezes in intestinal fluid of polar fishes. *Journal of Experimental Biology*, **98**, 429–38.

O'Grady, S.M., Schrag, J.D., Raymond, J.A. & DeVries, A.L. (1982*b*). Comparison of antifreeze glycopeptides from Arctic and Antarctic fishes. *Journal of Experimental Zoology*, **224**, 177–85.

Orci, L., Ravazzola, M.-J., Storch, Anderson, R.G.W., Vassalli, J.-D. & Perrelet, A. (1987). Proteolytic maturation of insulin is a post-Golgi event which occurs in acidifying clathrin-coated secretory vesicles. *Cell*, **49**, 865–8.

Petzel, D.H., Reisman, M.M. & DeVries, A.L. (1980). Seasonal variation of antifreeze peptide in the winter founder, *Pseudopleuronectes americanus*. *Journal of Experimental Zoology*, **211**, 63–9.

Prosser, C.L. (1973). Water: osmotic balance; hormonal regulation. In *Comparative Animal Physiology*, pp. 1–78. Philadelphia: W.B. Saunders.

Raymond, J.A. & DeVries, A.L. (1977). Absorption inhibition as a mechanism of freezing resistance in polar fishes. *Proceedings of the National Academy of Sciences, USA*, **74**, 2589–93.

Raymond, J.A., Radding, W. & DeVries, A.L. (1977). Circular dichroism of protein and glycoprotein fish antifreeze. *Biopolymers*, **16**, 2575–8.

Raymond, J.A., Wilson, P.W. & DeVries, A.L. (1989). Inhibition of growth of non-basal planes in ice by fish antifreeze. *Proceedings of the National Academy of Sciences, USA*, **86**, 881–5.

Schrag, J.D., O'Grady, S.M. & DeVries, A.L. (1982). Relationship of amino acid composition and molecular weight of antifreeze glycopeptides to non-colligative freezing point depression. *Biochimica et Biophysica Acta*, **717**, 322–6.

Schrag, J.D., Cheng, C.-H.C., Panico, M., Morris, H.R. & DeVries, A.L. (1987). Primary and secondary structure of antifreeze peptides from arctic and antarctic zoarcid fishes. *Biochimica et Biophysica Acta*, **915**, 357–70.

Scott, G.K., Hew, C.L. & Davies, P.L. (1985). Antifreeze protein genes are tandemly linked and clustered in the genome of the winter flounder. *Proceedings of the National Academy of Sciences, USA*, **82**, 2613–7.

Scott, G.K., Davies, P.L., Shears, M.A. & Fletcher, G.L. (1987). Structural variations in the alanine-rich antifreeze proteins of the Pleuronectinae. *European Journal of Biochemistry*, **168**, 629–33.

Scott, C.K., Hayes, P.H., Fletcher, G.L. & Davies, P.L. (1988). Wolffish antifreeze protein genes are primarily organized as tandem repeats that each contain two genes in inverted orientation. *Molecular and Cellular Biology*, **8**, 3670–5.

Shier, W.T., Lin, Y. & DeVries, A.L. (1972). Structure and mode of action of glycoproteins from an antarctic fish. *Biochimica et Biophysica Acta*, **263**, 406–13.

Shier, W.T., Lin, Y. & DeVries, A.L. (1975). Structure of the carbohydrate of antifreeze glycoproteins from an antarctic fish. *FEBS Letters*, **54**, 135–8.

Sonnichesen, F.D., Sykes, B.D., Chao, H. & Davies, P.L. (1993). The nonhelical structure of antifreeze protein type III. *Science*, **259**, 1154–7.

Van Voorhies, W.V., Raymond, J.A. & DeVries, A.L. (1978). Glycoproteins as biological antifreeze agents in the cod, *Gadus ogac*. *Physiological Zoology*, **51**, 347–53.

Wang, X., DeVries, A.L. & Cheng, C.-H.C. (1995*a*). Antifreeze peptide heterogeneity in an antarctic eel pout includes an unusually large major variant composed of two 7 kDa type III AFPs linked in tandem. *Biochimica et Biophysica Acta*, **1247**, 163–72.

Wang, X., DeVries, A.L. & Cheng, C.-H.C. (1995*b*). Genomic basis of antifreeze peptide heterogeneity and abundance in an antarctic eel pout–gene structures and organization. *Molecular Marine Biology and Biotechnology*, **4**, 135–47.

P.E. TIKU, A.Y. GRACEY and A.R. COSSINS

Cold–inducible gene transcription: Δ^9-desaturases and the adaptive control of membrane lipid composition

Introduction

An important response of animal tissues to environmental cold is to increase the proportion of unsaturated fatty acids in the membrane phospholipids. This response leads to a decrease in the structural order of the hydrocarbon interior of membranes and thereby offsets the ordering effects of cold (Cossins, 1994; Hazel & Williams, 1990). It has been termed 'homeoviscous adaptation' (HA) in recognition of its homeostatic and adaptive significance (Sinesky, 1974), and has been widely observed not only in animals but also in plants and micro-organisms during both phenotypic and genotypic responses to altered environmental temperature. HA appears to play one of two important adaptive roles. Firstly, it may ensure that the fine physical structure of cellular membranes, and hence those functions which are dependent on structure, are preserved when growth temperature is altered despite the intrinsic temperature dependance of these phenomena. This is a capacity adaptation in the sense of Precht (Cossins & Bowler, 1987). Secondly, it may alter the sensitivity of cell, tissue and whole-organism function to disruption at extreme temperatures, thereby improving the match between thermal resistance and environmental temperature. This is a resistance adaptation.

This general scheme is now supported by a considerable amount of evidence linking changes in lipid unsaturation during thermal acclimation to changes in membrane physical structure and, in a few cases, to changes in membrane-associated function (Cossins, 1994). The linkages have been established both in direction and in time course. However, it should be recognized that they constitute correlations, and despite their plausibility do not constitute unequivocal proof of a causal relationship.

There are at least two reasons to question the tacit assumption of a causal link between membrane lipid unsaturation, a compensated membrane structure and adapted membrane function. The first is that the compositional response to temperature is complex and that, whilst some parts of the response may have adaptive significance, others may not. There may be dozens of different fatty acids, all of which might change in proportion. There may be changes in the positional disposition on the *sn-1* and *sn-2* position, in the distribution of phospholipid headgroups between inner and outer monolayer, between different domains or between different membrane types within the cell. Finally, there may be changes in cholesterol content and distribution. There is every reason to believe that this complexity in composition has equally complex effects upon biomembrane physical structure, including, for example, changes to the gradient of hydrocarbon segmental order and flexibility with increasing depth into the bilayer, to the miscibility of laterally separated micro domains of more ordered and more disordered lipids in the plane of the membrane and in the monolayer specificity of different types of phospholipid. This complexity makes it difficult to establish a precise linkage between any specific compositional changes and its effect upon membrane physical structure. Are we to believe that all aspects of the compositional response contribute to the observed temperature compensation of membrane structure? Or are some aspects unimportant, or at least of lesser importance than other aspects? Are some compositional responses linked to specific aspects of the structural response? Because each aspect of the response forms an integral part of a much larger ensemble which *in vivo* is expressed in its entirety, the structural significance of any specific aspect of the response cannot be addressed by the comparative approach.

A second important limitation of current approaches is that they do not allow the assessment of the adaptive significance to the whole organism of specific aspects of the overall homeoviscous response, except by inference. It is one thing to demonstrate, using reconstituted membranes, that a compositional modification might lead to a specific change in physical structure, but it is quite another thing to assume that this specific modification is important to Darwinian fitness at the whole organism level. The basic problem is that in studies of homeoviscous adaptation in animals we are presented with a physiological response which is not subject to firm experimental control. Specific aspects of the response cannot be abolished or enhanced such that the effects upon function and fitness may be assessed.

Modern molecular and genetic techniques provide powerful new approaches for investigating these links and undoubtedly these techniques will progressively replace the conventional analytical and biophysical approaches. These methods allow the expression of specific proteins to be controlled, either by abolition or by overexpression. This can be achieved *in vitro* and increasingly *in vivo* by gene manipulation and not only can the structural and functional significance of that enzyme be assessed in isolated membranes but in certain circumstances the fitness attributes of the manipulated phenotype can be compared with individuals expressing the normal, wild-type response. It is the specificity of control that permits a more refined and discrete analysis of the causal links between composition, structure and function. This specificity also allows questions to be asked regarding the adaptive significance of specific attributes of altered enzyme expression. Genetic manipulation can allow the performance and competitive fitness of organisms or cultured cells to be assessed in relation to the role of specific genes and proteins for which they code.

Adoption of this approach requires firstly the identification of important controlling enzymes in the compositional homeoviscous response and an understanding of how expression of that enzyme is altered. Because homeoviscous adaptation has been linked with changes in acyl group saturation most attention has been paid to the control of three enzymatic activities, the elongases, the desaturases and the acyltransferases. As we shall see, the activity of the Δ^9-desaturase is frequently increased in cold-acclimated organisms and this enzyme has become the focus of much interest in understanding the mechanisms of membrane adaptation. However, studies of this sort have a more general significance in that HA is perhaps the most clearcut example of membrane homeostasis, a phenomenon common to all living organisms. Understanding membrane adaptation to temperature offers insights into how cells generally control the composition and physical structure of cellular membranes.

Lipid desaturases and homeoviscous adaptation

There is now strong evidence that the introduction of the first unsaturation bond has by far the greatest effect on membrane physical properties and successive bonds have progressively smaller effects (Coolbear, Berde & Keogh, 1983; Stubbs *et al.*, 1981). It is also clear that incorporation of the first double bond at the central 9–10 position from the carboxyl terminus has maximal effects on lipid physical properties (Barton & Gunstone, 1975). These two observations indicate that membrane physical properties can be controlled most effectively, at least over a certain range, by regulating the proportion of saturated fatty acids compared to unsaturated fatty acids and this can be achieved by modifying the activity of the enzyme, the Δ^9-desaturase, which incorporates the first double bond into stearic or palmitic acid (Macartney, Maresca & Cossins, 1994).

Desaturases are present in virtually all living organisms. They fall into three main groups depending on the substrate being a CoA-derivatized acyl group, an acyl carrier protein linked lipid as in plants (McKeon & Stumpf, 1982) or an acyl group linked to a complex glycerophospholipid as found in plants and cyanobacteria (Jaworski, 1987; Wada *et al.*, 1993). The hepatic Δ^9-desaturase in animal cells is of the acyl-CoA type. It is bound to the endoplasmic reticulum in animal and yeast cells (Enoch, Catala & Strittmatter, 1976; Strittmatter *et al.*, 1974) where it forms the terminal component of a multi-component system which includes cytochrome b_5 and the NADH-dependent cytochrome b_5 reductase. This complex catalyses the insertion of a double bond at the 9–10 position in a variety of saturated fatty acid-CoA substrates (mainly stearoyl- and palmitoyl-CoA) and accounts for all *de novo* synthesized *cis*-unsaturated fatty acids. It is generally believed that the rate of the overall reaction is limited by the terminal component, the desaturase, and it is this component that shows changes in expression with experimental treatment, be it dietary or thermal.

Adjustments to thermal stress by altering desaturase activity have been observed in all groups of living organisms. The early and classical studies by Fulco on the micro-organism, *Bacillus megaterium*, demonstrated that cooling of cultures led to a 'hyperinduction' of desaturase activity above the levels observed in the steady state at any temperature (Fujii & Fulco, 1977). Cold-increased desaturase activity has also been confirmed for the cyanobacteria *Anabaena variabilis* and *Synechocystis* (Sato & Murata, 1981; Wada, Gombos & Murata, 1990). The thermo-tolerant strain, NT-1, of the protozoan *Tetrahymena pyriformis* shows an increased Δ^9-desaturase activity which occurred within 2 hours, following the shift in temperature from 39.5 to 15 °C (Fukushima *et al.*, 1976; Nozawa *et al.*, 1974). The cold-induced change in fatty acid composition of membrane phospholipids in animals is more complex and it is likely that coordinated changes in the expression of several and probably many enzymes is involved. However, there is one particularly clear-cut compositional response which is perhaps the most striking example of temperature-induced enzymatic induction in higher animals. Schünke and Wodtke (Schünke & Wodtke, 1983) showed that the progressive cooling of carp over a 3-day period led to a pronounced 14–32-fold increase in Δ^9-desaturase activity. Activity increased rapidly to peak values after day 4 then declined only to increase again on day 10. Both peaks were associated with shifts in the level of unsaturated fatty acids in phospholipids of the endoplasmic reticulum though only the first was linked with changes in membrane physical order as detected by fluorescence polarization spectroscopy (Wodtke & Cossins, 1991).

It is worth emphasizing that, in most of these cases, the cold induction of Δ^9-desaturase activity is transient, such that activity is low both before and after the temperature shift. Evidently, a continued high level of expression is unnecessary at both high and low temperatures and we presume that, once the necessary changes in lipid composition have been introduced, expression is down-regulated. This supports the notion of an active feedback regulation of desaturase activity and the highly dynamic nature of desaturase turnover and of the control process. Thus, it is absolutely necessary to define changes in activity during the early phase of adjustment rather than at the steady state, i.e. within hours and days rather than weeks and months. Another important point in the studies by Wodtke is that cold induction was only observed in carp fed a diet rich in unsaturated fatty acids (Wodtke, 1986; Wodtke, Teichert & Konig, 1986). Warm acclimated carp fed a more saturated diet possessed appreciable desaturase activities which showed no great change on cold exposure. Evidently, desaturase expression is sensitive to dietary influences, and any experimental investigation must be designed with this in mind.

Mechanisms of desaturase induction

There are two obvious and distinct ways of modifying cellular desaturase activity; by changes to the number of active enzymes or by changes to their molar activity (Macartney *et al.*, 1994). In discussing the processes which lead to desaturase induction, we distinguish the proximal mechanism producing the change in desaturase activity (e.g. translational or transcriptional up-regulation) from the ultimate controlling process which regulates the proximate mechanism. The ultimate mechanism presumably responds to a change either in temperature *per se* or in some aspect of the compositional or physical condition of the membrane. A major issue here is the probable existence and nature of a feedback element which links the controlled variable with the controlling process. As already indicated, control appears to be complex in that desaturase induction is transient and this suggests that desaturase activity is subject to fine control through some negative feedback. Understanding how this feedback works is an essential part of understanding not only the mechanism but also the potential for manipulation.

Viscotropic regulation

An early hypothesis was that the desaturase was affected unconventionally by lipid 'viscosity', such that increased membrane order directly caused an increased activity and vice versa (Kasai *et al.*, 1976; Martin *et al.*, 1976; Thompson & Nozawa, 1977). This idea initially proved attractive because it neatly incorporated a feedback element in the localized lipid microenvironment of the enzyme and was supported by experiments using fluidizing drugs or cholesterol (Garda & Brenner, 1984, 1985). It had several other attractive features. Firstly, it identified the controlled variable as that aspect of lipid order which influenced desaturase activity. Secondly, it suggested that different stimuli (temperature, lipid diet, hydrostatic pressure, fluidizing drugs) would act through a common mechanism, that is their effects upon that aspect of lipid structure that influenced desaturase activity. Thirdly, it suggested that the change in desaturase activity following a change in temperature was immediate and the time for the resulting change in lipid saturation and lipid physical order depended upon the rates of lipid turnover.

Despite some supportive evidence, the prediction that desaturase activity would be increased by the ordering effects of a decreased temperature has never been verified (Schünke & Wodtke, 1983). Moreover, it is difficult to rationalize how a negative temperature coefficient might be achieved, and this provides powerful evidence against the hypothesis at least as applied to thermal adaptation. Unfortunately, without measurements of the number of active desaturases, it is not possible to be confident that rapid changes in desaturase activity are caused by changes in molar activity rather than through changes in the number of active units.

Altered protein turnover by degradative processes

The number of desaturases present may be controlled in the longer term through altered turnover of the enzyme and this can be achieved by changes in the rates of either synthesis or degradation. Control by altered degradation is a comparatively unexplored mechanism for desaturases and for membrane-bound enzymes generally. However, changes in the expression of the rate-limiting enzyme in cholesterol biosynthesis, the HMG-CoA reductase, offers relevant insights into the possibilities of feedback regulation of membrane bound enzymes. The effective degradation of HMG CoA reductase requires that the cytoplasmic domain of the enzyme must be attached to the membrane-bound domain (Gil *et al.*, 1985), possibly because the relevant protease is membrane associated, and access to this protease is limited by lateral diffusion. Cold will reduce the lateral mobility of endoplasmic reticulum proteins, perhaps drastically, and this will impede their delivery to the putative protease. The resulting greater decrease in degradative rate relative to the rate of synthesis would then lead to increased steady-state levels of that protein. This scheme constitutes a negative feedback in that increased lipid order in the cold itself reduces degradative rates, and the resulting change in desaturase expression brings about a compensatory change in the saturation of the fatty acid pool and ultimately in the saturation of the endoplasmic reticulum phospholipids.

Altered desaturase synthesis

Rates of desaturase synthesis may be limited by transcriptional, post-transcriptional and translational mechanisms. In *B. megaterium*, the regulation of Δ^9-desaturase activity is achieved in two ways (Fujii & Fulco, 1977); firstly, by the regulation of gene transcription by means of a temperature-sensitive modulator protein and, secondly, by changes in protein thermostability and hence degradation. Recent work from Murata's group has shown convincingly that cold exposure of the cyanobacterium *Synechocystis* dramatically increases the yield of mRNA coding for a desaturase gene presumably by means of induced transcription (Los *et al.*, 1993). They have subsequently shown that the catalytic hydrogenation of plasma membrane lipids induces an isothermal production of mRNA and that hydrogenation of membrane lipids sensitizes cells to cooling (Vigh *et al.*, 1993). This important work provides the first evidence of a transcriptional control of the desaturase gene by means of a feedback element located at the plasma membrane (Maresca & Cossins, 1994). Whether this is linked to physical structure or 'fluidity', or to the level of lipid saturation, remains to be seen. Finally, a role for transcription in controlling desaturase expression is indicated by a number of recent studies of dietary-treated rats and mice in that mRNA abundance is increased. However, the roles of induced transcription or altered transcript turnover have not been separated explicitly.

Identifying possible transduction mechanisms which allow for feedback of a signal from the membrane to the nucleus is perhaps the least understood aspect of the entire response. Some idea of potential mechanisms by which transcription may be regulated is provided by 3T3-L1 preadipocytes. Differentiation of these cells leads to activation of one of the two Δ^9-desaturase genes (*SCD2*) (Kaestner *et al.*, 1989;

Ntambi *et al.*, 1988); and transfection experiments using the *SCD2* promoter with a reporter gene have identified a repressor element, termed *PRE* (Swick & Lane, 1992). This element is located approximately 400 bp upstream of the promoter but is inactive in adipocytes. A 58 kD *PRE*-binding protein has been identified by gel retardation experiments. Thus, desaturase induction might involve the activation of a transcriptional factor, either negatively or positively, by a mechanism that is controlled by some aspect of membrane physical condition.

Altered desaturase expression during cold acclimation of carp

We have recently implemented a molecular biology approach to understanding the mechanism underlying the cold-induced increase in desaturase activity using the carp liver system (Tiku *et al.*, 1996). Recent developments in the molecular analysis of mammalian Δ^9-desaturases have provided the appropriate immunological and genetic tools for the analysis of desaturase expression (Strittmatter *et al.*, 1988; Thiede, Ozols & Strittmatter, 1986; Thiede & Strittmatter, 1985) given an adequate degree of homology between rat and carp sequences.

The cDNA for the rat desaturase encodes a protein of 358 amino acids with a predicted molecular weight of 41.4 kD (Strittmatter *et al.*, 1988; Thiede & Strittmatter, 1985). As already indicated, two Δ^9-desaturase genes (*SCD1* and *SCD2*) have been identified in mouse and rat which encode two isoforms (Kaestner *et al.*, 1989; Mihara, 1990; Ntambi *et al.*, 1988) displaying a tissue-specific distribution. Mouse *SCD1* exhibits 92% homology with rat *SCD1* at the amino acid level whilst *SCD2* shows 86% and 87% homology to rat and mouse *SCD1*, respectively. The Δ^9-desaturase gene (*OLE1*) of the yeast *Saccharomyces cerevisiae* codes for a protein of 510 amino acids with a calculated mass of 58.4 kD (Stukey, McDonough & Martin, 1989, 1990). At the amino acid level it shows 36% identity and 60% similarity with the rat *SCD* and there are three highly conserved regions including one with 11 out of 12 perfect residue match. Significantly the yeast protein can be functionally replaced by the rat *SCD* gene (Stukey *et al.*, 1990) and a similar experiment has been performed to complement a desaturase mutation in tobacco plants (Grayburn, Collins & Hildebrand, 1992). Thus, Δ^9-desaturase structure and function appears to be conserved across a wide range of organisms and this encourages an approach based on heterologous cDNA screening.

We have attempted to raise PCR products from carp liver cDNA using primers corresponding to the conserved segments in the rat liver and yeast gene Δ^9-desaturase cDNA. This approach proved unsuccessful, possibly because the degeneracy built into the primers reduced the stringency for selective amplification of desaturase to levels which prevented production of any useful DNA. An alternative approach was based on screening a commercial carp liver cDNA library with the rat cDNA. Early experiments indicated that the homology at the nucleotide level between carp and rat was too low to allow detection of a desaturase mRNA by stringent Northern analysis. However, subsequent work on isolated carp hepatocytes indicated that continued long-term culture led to a large isothermal desaturase induction (Macartney, Tiku &

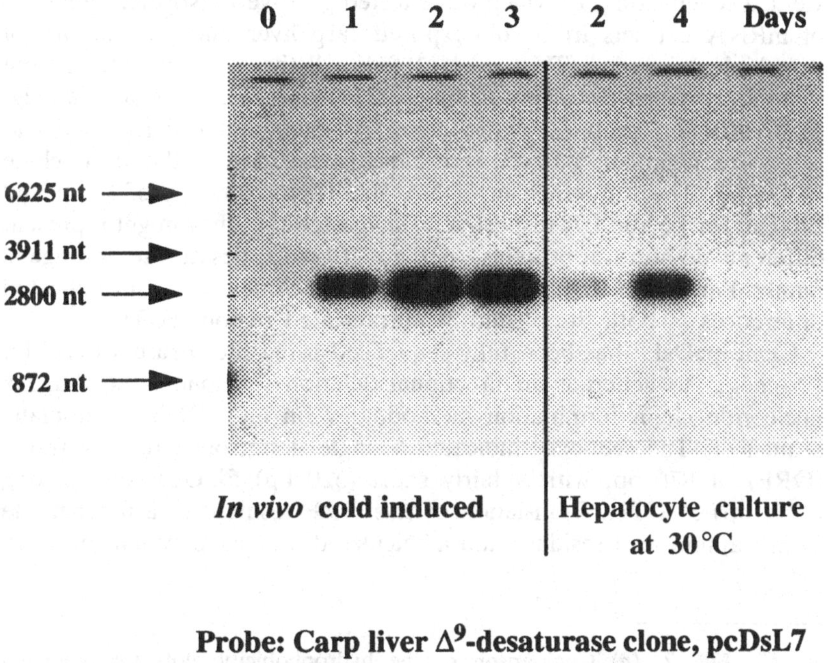

Fig. 1. Northern analysis of hepatic RNA samples using a probe derived from the carp liver desaturase clone, pcDsL7. Samples were extracted from carp maintained for several months at 30 °C (day 0) or at the indicated period following initiation of transfer to 10 °C. Extracts from hepatocyte culture were obtained after the indicated times following initiation of culture of cells from 30 °C acclimated fish at 30 °C.

Cossins, 1995) and this was related to the increasing quantities of a transcript which could be detected on Northern analysis using the rat cDNA as probe (see, for example, Fig. 1). The apparently large quantities of mRNA on hepatocyte induction together with very long exposure periods for autoradiography allowed refinement of the conditions for selective heterologous hybridization with moderate stringency.

The level of this transcript revealed by Northern analysis mirrored the increasing enzymatic activity of the desaturase during hepatocyte culture as well as the change in membrane phospholipid fatty acid composition and this encouraged a belief that the rat cDNA probe could be useful in isolating carp desaturase sequences. Successive heterologous screening of a commercial carp hepatocyte cDNA library yielded seven clones of which three tested positive in Northern analyses of mRNA extracts from cold-exposed carp liver under conditions of very high stringency (Tiku *et al.*, 1996). All three clones detected the same sized transcript (approximately 2700 bp; see, for example, Fig. 1). Restriction analyses of the clones revealed that two (pcDsL5 and L6) were identical, but showed some similarities to the third clone (pcDsL7). The difference in cDNA size (approximately 5500 bp and 2700 bp for L6 and L7, respectively) suggested that they might represent different desaturase proteins, perhaps from a desaturase multigene family. Partial DNA sequencing of these two clones has so far indicated differences in both the 5′ and 3′ untranslated regions (UTRs).

Clone pcDsL7 has been fully sequenced using the 'Erase-a-Base' kit (Promega) to generate unidirectional deletion subclones, and dideoxy nucleotide chain termination methodology on an ABI 373 automatic sequencer. The sequence indicated a single, long, open reading frame (ORF) of 876 bp, with a fairly short (520 bp) 5′ UTR and a long (1271 bp) 3′ UTR. Translation of this ORF resulted in a polypeptide of 292 amino acid residues and a calculated molecular weight of 33.75

Fig. 2. (*a*) Comparison of the hydrophobicity plots for predicted Δ^9-desaturase from carp, rat, tick and yeast. The published nucleotide sequences from the latter three were obtained from GenBank. Kyte–Doolittle plots with 9 amino acid residue windows were generated with Lasergene software from DNASTAR. (*b*) Comparison of the predicted amino acid sequences for Δ^9-desaturases of carp, rat, tick and yeast. Clustal alignment using a structural residue weight table (DNASTAR Lasergene software). The shaded areas indicate residues which match the consensus within one distance unit.

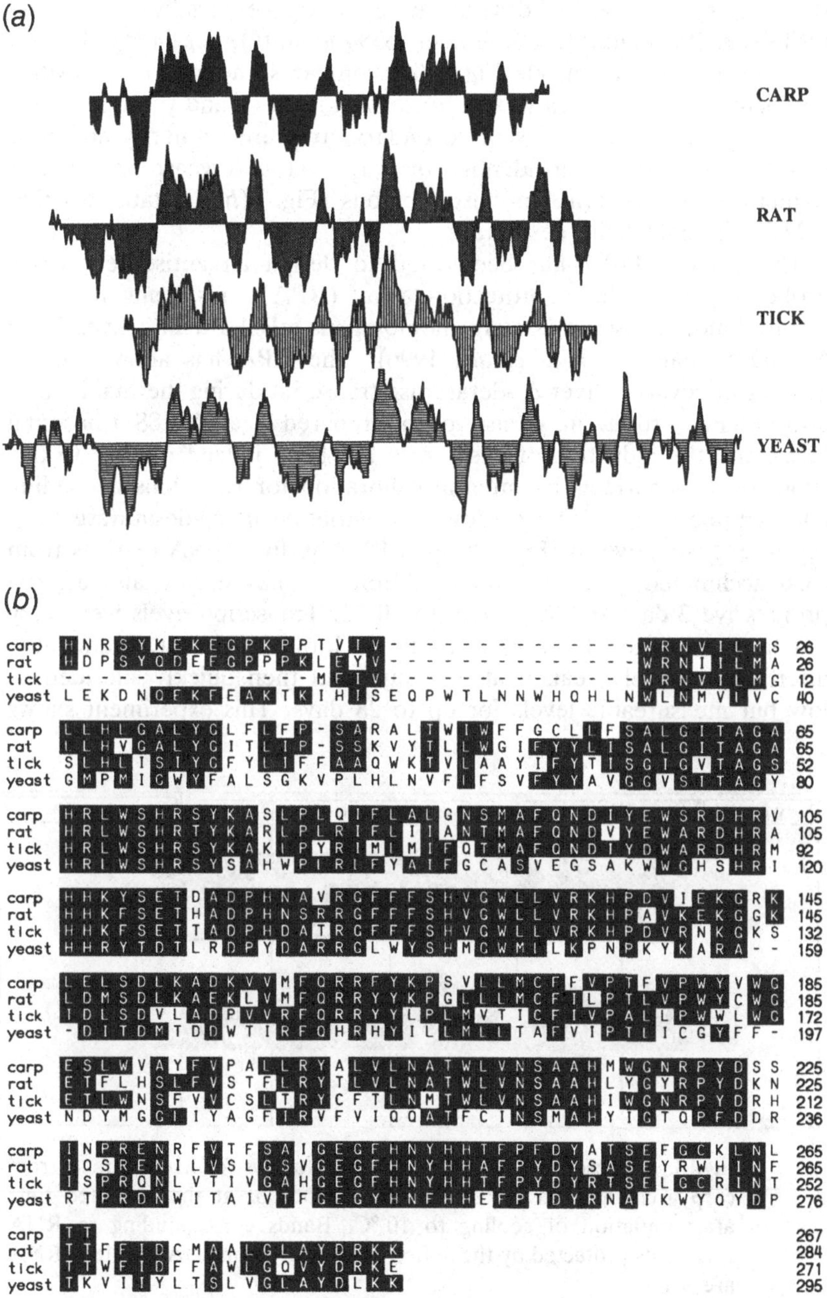
(a)
CARP
RAT
TICK
YEAST
(b)
carp H N R S Y K E K E G P K P P T V I V - - - - - - - - - - - - - - - W R N V I L M S 26
rat H D P S Y Q D E E G P P P K L E Y V - - - - - - - - - - - - - - - W R N I I L M A 26
tick - - - - - - - - - - - - - - M F I V - - - - - - - - - - - - - - - W R N V I L M G 12
yeast L E K D N Q E K F E A K T K I H I S E Q P W T L N N W H Q H L N W L N M V I V C 40
carp L L H L G A L Y G L F L F P - S A R A L T W I W F F G C L L F S A L G I T A G A 65
rat I L H V G A L Y G I T L I P - S S K V Y T L L W G I F Y Y L I S A L G I T A G A 65
tick S L H L I S I Y G F Y L I F F A A Q W K T V L A A Y I F Y T I S G I G V T A G S 52
yeast G M P M I C W Y F A L S G K V P L H L N V F I F S V F Y Y A V G G V S I T A G Y 80
carp H R L W S H R S Y K A S L P L Q I F L A L G N S M A F Q N D I Y E W S R D H R V 105
rat H R L W S H R T Y K A R L P L R I F L I I A N T M A F Q N D V Y E W A R D H R A 105
tick H R L W S H R S Y K A K L P Y R I M I M I F Q T M A F Q N D I Y D W A R D H R M 92
yeast H R I W S H R S Y S A H W P I R I F Y A I F G C A S V E G S A K W W G H S H R I 120
carp H H K Y S E T D A D P H N A V R G F F F S H V G W L L V R K H P D V I E K G R K 145
rat H H K F S E T H A D P H N S R R G F F F S H V G W L L V R K H P A V K E K G G K 145
tick H H K F S E T T A D P H D A T R G F F F S H V G W L L V R K H P D V R N K G K S 132
yeast H H R Y T D T L R D P Y D A R R G L W Y S H M G W M I L K P N P K Y K A R A - - 159
carp L E L S D L K A D K V V M F Q R R F Y K P S V L L M C F F V P T F V P W Y V W G 185
rat L D M S D L K A E K L V M F Q R R Y Y K P G L L L M C F I L P T L V P W Y C W G 185
tick I D L S D V L A D P V V R F Q R R Y Y L P L M V T I C F I V P A L L P W L W G 172
yeast - D I T D M T D D W T I R F Q H R H Y I L L M L I T A F V I P T L I C G Y F F - 197
carp E S L W V A Y F V P A L L R Y A L V L N A T W L V N S A A H M W G N R P Y D S S 225
rat E T F L H S L F V S T F L R Y T L V L N A T W L V N S A A H L Y G Y R P Y D K N 225
tick E T L W N S F V V C S L T R Y C F T L N M T W L V N S A A H I W G N R P Y D R H 212
yeast N D Y M G G I I Y A G F I R V F V I Q Q A T F C I N S M A H Y I G T Q P F D D R 236
carp I N P R E N R F V T F S A I G E G F H N Y H H T F P F D Y A T S E F G C K L N L 265
rat I Q S R E N I L V S L G S V G E G F H N Y H H A F P Y D Y S A S E Y R W H I N F 265
tick I S P R Q N L V T I V G A H G E G F H N Y H H T F P Y D Y R T S E L G C R I N T 252
yeast R T P R D N W I T A I V T F G E G Y H N F H H E F P T D Y R N A I K W Y Q Y D P 276
carp T T 267
rat T T F F I D C M A A L G L A Y D R K K 284
tick T T W F I D F F A W L G Q V Y D R K E 271
yeast T K V I I Y L T S L V G L A Y D L K K 295

kD. The predicted protein showed a high degree of sequence similarity with other published Δ^9-desaturase sequences, especially those from rat liver (55% identity), mouse liver (53% identity), tick (47% identity) and yeast (20% identity). Fig. 2(*a*) compares the predicted hydrophobicity plots for the carp desaturase with rat, tick and yeast enzymes, revealing extensive regions of secondary structure similarity and this, together with the near identity of carp, rat and yeast amino acid sequences over several consensus regions (Fig. 2(*b*)), establishes that pcDsL7 is indeed the desaturase.

The clone pcDsL7 has been used to design an antisense mRNA probe for ribonuclease protection assay (RPA), this being a 50- to 100-fold more sensitive assay for homologous mRNA than conventional Northern analyses (Tiku *et al.*, 1996). The RPA has allowed us to investigate levels of liver Δ^9-desaturase transcript during thermal acclimation. RPA probes have also been prepared against 18S ribosomal RNA and the induction of desaturase has been judged relative to this (Fig. 3). This corrects by internal calibration for variations in loading and running of gels. Fig. 4 shows the variation in Δ^9-desaturase transcript levels relative to 18S ribosomal RNA in liver RNA extracts from carp acclimated to 30 °C and at different times during and after a progressive 3-day cooling regime to 10 °C. Transcript levels were very low in warm-acclimated carp and after 24 h of cooling but accumulated after 48 h. Levels peaked at 2–3 days and then quickly subsided to low but measureable levels for up to 28 days. This experiment shows

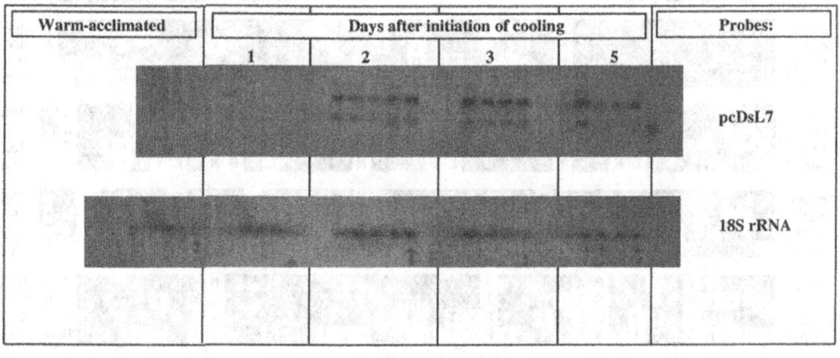

Fig. 3. Representative RNase protection assays for liver extracts from carp acclimated to 30 °C for long periods or at the indicated time after initiation of cooling to 10 °C. Bands corresponding to RNA fragments protected by the antisense desaturase mRNA and 18S rRNA are shown.

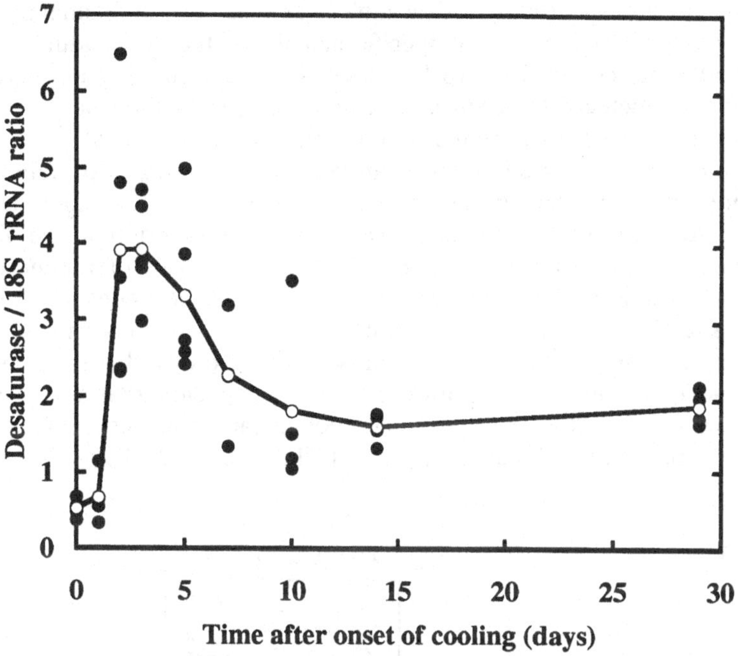

Fig. 4. The effect of cold acclimation of carp upon the levels of hepatic Δ^9-desaturase mRNA transcript. Total RNA was extracted from individual carp subjected to indicated periods of the cooling regime from 30 °C down to 10 °C, as previously described (Schünke & Wodtke, 1983). Desaturase mRNA levels have been determined by densitometric analysis of RPA autoradiographs. Values were internally normalized to that determined for the 18S rRNA levels. Filled symbols represent data from individual animals (4–5) and unfilled symbols represent mean values.

that, although cooling led to a dramatic increase in transcript levels, the effect was transient. Evidently, the change in saturation requires only a transient increase in expression and the low levels are entirely sufficient to maintain the *status quo* on long-term exposure to cold. Exactly how the increase in transcript levels is brought about is not entirely clear. It could be due either to activation of transcription or to a change in transcript turnover due to a cold-induced reduction in mRNA degradation. Preliminary results from nuclear run-on experiments indicate that the former is involved.

Recently, we have found that a polyclonal antibody raised against purified rat *SCD* by Strittmatter (Thiede & Strittmatter, 1985) reacts against a carp protein with a molecular mass of approximately 33 kD (Fig. 5). The increase in specific activity of the Δ^9-desaturase during isothermal culture of carp hepatocytes was matched by an increase in immuno-detectable desaturase protein, suggesting that the induction of specific activity was linked to changes in turnover of the protein, rather than to some late post-translational processing of protein already present in the membrane. However, in more recent experiments on *in vivo* thermally acclimated carp we have found that the liver of warm-acclimated carp possesses considerable levels of immunodetectable desaturase and that this increases only by a modest 30% on day 5 after initiation of cooling. Thus, whilst the liver of warm-acclimated carp displayed little or no desaturase activity, they did possess considerable quantities of desaturase protein. It appears that these animals possess a latent enzyme and that cooling causes its activation, possibly by a post-translational mechanism (Tiku *et al.*, 1996).

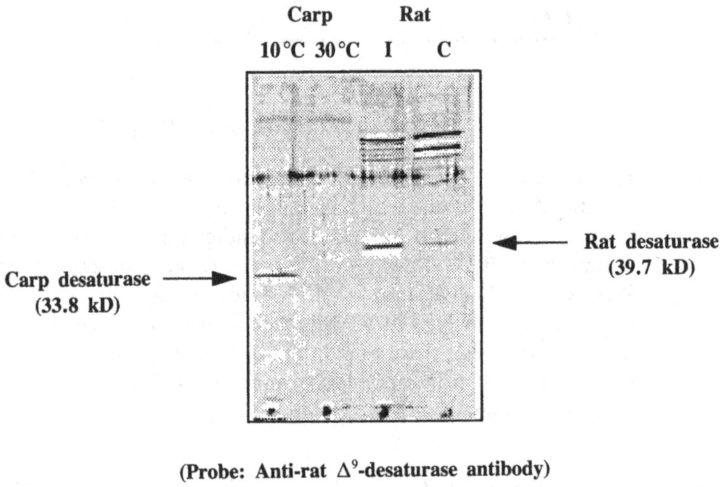

Fig. 5. Western immunoblot of carp and rat liver microsomal proteins using anti-rat desaturase antibody described by Strittmatter (Thiede & Strittmatter, 1985; Strittmatter *et al.*, 1988). The lanes marked **C** and **I** represent rat extracts prepared from animals fed a normal rat chow (Control) or after a starvation-refeed strategy (Induced), respectively, in order to induce the rat desaturase. The carp lanes were prepared from fish held at 30 °C for a long period, or 5 days after transfer of warm-acclimated fish to 10 °C. Note the difference in molecular weight of the immunopositive bands from carp and rat matches that was predicted from a knowledge of their nucleotide sequences.

We conclude that cold induction of the Δ^9-desaturase is due to two events. Firstly, a pre-existing desaturase is activated during the early stages of cold transfer, possibly by a post-translational modification. Secondly, the levels of desaturase transcript are increased possibly due to induced transcription and this presumably leads to an enhanced rate of desaturase synthesis, as evidenced by the increase in immunodetectable protein after 5 days of cooling. Both events correlate in time course and in direction with the change in lipid saturation and membrane lipid order. Thus control of desaturase expression is complex, at least under the conditions applied in this experiment. The relationship between the various component responses is not clear. It may be that the compound response is made up of components which are recruited at different phases of the progressive cooling experienced by the fish. Thus, the post-translational effect might be an early response to moderate cooling whilst the transcriptional response, and the consequent increase in desaturase levels, might occur only at lower temperature. As regards the possible feedback element, our results indicate that variations between individual fish in mRNA levels is correlated with the fatty acid composition of the microsomal lipids; individuals with more saturated microsomal phospholipids exhibit higher levels of desaturase mRNA. This preliminary evidence provides some positive evidence that the presumed transcriptional activation is linked to the saturation or physical condition of endoplasmic reticulum membranes.

Future developments

Identification of the desaturase as an important, if not critical, component in homeoviscous adaptation and the isolation of cloned genes and production of antibodies against the Δ^9-desaturase allows the mechanism and adaptive significance of homeoviscous adaptation to be addressed in a more direct manner. Specifically, we envisage progress to be made in three directions.

Mechanisms of cold-induced transcriptional activation?

We have demonstrated an increase in transcript levels during the early stages of homeoviscous adaptation. As already indicated, a particularly interesting question is how desaturase gene transcription is controlled in response to cooling? Whilst we recognize that induction might be due to changes in RNA turnover, a more likely possibility is that induced transcription is involved. In general, transcription is controlled by a series of DNA-binding proteins which by interacting with specific regulatory DNA sequences influence the transcriptional activity of RNA polymerase II. The number of DNA-binding proteins may be large and their effects and interactions in promoting transcription may be complex, as in the induction of heat shock proteins (Morimoto, Tissieres & Georgopoulos, 1994). Nevertheless, progress in understanding how cold exposure modulates the activity of DNA-binding proteins requires the identification and characterization of the regulatory proteins. We have termed the putative promoter–enhancer complex of carp liver the 'cold-inducible promoter' system or CIP, and we are actively seeking to identify its position in relation to the coding sequence for the Δ^9-desaturase gene.

Another important question relates to the mechanism which induces the transcriptional up-regulation. In other words, what is the regulated variable and what is the temperature sensor in what may be a long and complex feedback pathway? The work from the groups of Murata and Vigh (Horvath *et al.*, 1991; Vigh *et al.*, 1993) on the Δ^{12}-desaturase implies that the sensor responds to changes in the physical properties of the membrane. This is a very important observation and repeating this experiment in a vertebrate cell line would offer direct evidence of a feedback linked to membrane fluidity itself, as previously suggested (Maresca & Cossins, 1994). Exactly how a physical property of the plasma membrane influences transcriptional activity in the nucleus is difficult to explain but it may involve a phosphorylation cascade, one of whose components binds to a membrane in a manner that depends upon lipid order.

Induced expression of other lipid biosynthetic enzymes?

HA requires a very complex restructuring of the membrane lipids throughout the cell. These changes result from the concerted activity of various enzymes and not just through changes in desaturase expression. Thus, Hazel and colleagues (Hazel & Landrey, 1988*a*, *b*; Hazel *et al.*, 1991) have shown a complex overlapping time course of changes in various types of compositional adjustment (molecular species distribution, phospholipid headgroup, monoenes, polyenes, cholesterol, etc.), and we would expect that these were mediated by the altered expression of a suite of enzymes. The approach adopted for desaturase is capable of being extended to other types of enzyme especially as it is likely that they are controlled by the same regulatory factors that bind at the desaturase CIP. Dietary effects on various genes appear to be regulated by 'fat specific elements' and these might also be involved in the CIP (Mihara, 1990; Ntambi *et al.*, 1988). Recent advances in molecular techniques, including differential display reverse transcriptase PCR facilitate the cloning of genes whose transcription is induced by experimental treatment, manipulation or disease.

A causal role of desaturases in HA?

An ability to control the expression of the desaturase during cold transitions or even at normal temperatures allows a much more direct analysis of the role of this enzyme in controlling membrane lipid composition and 'fluidity'. Thus, overexpression of the desaturase on cold transfer should produce a greater HA and shift in lipid composition. Similarly, induced expression of the desaturase under the control of an inducible promoter should increase the proportion of monoenes with some effects on 'fluidity'. Conversely, abolition of desaturase induction either by gene-targeted mutation or by the introduction of antisense transcripts should lead to reduced or totally absent HA. Finally, the production of transgenic carp allows questions relating to the link between HA and cold tolerance of the whole animal to be addressed. Thus, incorporating the desaturase coding sequence under the control either of an inducible promoter, such as metallothionein promoter, or of the CIP itself to give multiple copies should lead to an enhanced homeoviscous response. Conversely, introduction of antisense desaturase or of a mutated desaturase might abolish enzyme induction, the homeoviscous response and acquired cold tolerance. Indeed, the use of chimaeric or mosaic individuals allows investigation of the temperature tolerance of specific tissues for which the untransformed controls are provided within the same individual.

Genetic manipulation of the desaturase gene in cyanobacteria has offered particularly interesting insights into the function of desaturases in homeoviscous adaptation and acquired thermotolerance. The unsaturation of fatty acids in membrane lipids can be modified by genetic manipulation of the various desaturase genes in *Synechocystis* strain PCC6803 (Sakamoto *et al.*, 1994; Wada *et al.*, 1990) and this influences growth and chilling tolerance (Gombos, Wada & Murata, 1991, 1992, 1994). Elimination of desaturation at the Δ^{12} position of fatty acids had a deleterious effect on the growth of this micro-organism at 22 °C (cf. normal growth temperature of 34 °C), indicating the importance of incorporation of unsaturated fatty acids at lowered growth temperatures. Furthermore, transformation of *Synechocystis sp.* PCC7942 with the *desA* gene, enhanced chill tolerance compared to the wild-type cells, clearly demonstrating the role of the gene in chilling tolerance and the maintenance of photosynthetic ability.

Genetic manipulations of higher plants have also elegantly demonstrated a similar critical role of lipid biosynthetic enzymes in cold tolerance (Ishizaki *et al.*, 1988; Murata *et al.*, 1992; Nishida *et al.*, 1993). Thus, the introduction of glycerol-3-phosphate acyltransferase cDNA from squash (which is chill sensitive) or from *Arabidopsis* (which is chill-resistant) into tobacco (which has intermediate chill-sensitivity) resulted in altered chilling sensitivity and resistance, respectively. In the former case, the level of unsaturated fatty acids in phosphoglycerolipids fell significantly and chilling sensitivity was markedly increased, whilst in the latter there was an increase in both the unsaturation of fatty acids in phosphoglycerolipids and chilling tolerance.

References

Barton, P.G. & Gunstone, F.D. (1975). Hydrocarbon chain packing and molecular motion in phospholipid bilayers formed from unsaturated lecithins. *Journal of Biological Chemistry*, **250**, 4470–6.

Coolbear, K.P., Berde, C.P. & Keogh, K.M.W. (1983). Gel to liquid-crystalline phase transitions of aqueous dispersions of polyunsaturated mixed acid phosphatidylcholines. *Biochemistry*, **22**, 1466–73.

Cossins, A.R. (1994). Homeoviscous adaptation and its functional significance. In *Temperature Adaptation of Biological Membranes*, ed. A.R. Cossins, pp. 63–75. London: Portland Press.

Cossins, A.R. & Bowler, K. (1987). *The Temperature Biology of Animals*, pp. 339. London: Chapman & Hall.

Enoch, H.G., Catala, A. & Strittmatter, P. (1976). Mechanisms of rat liver microsomal stearyl-CoA desaturase. Studies of the substrate specificity, enzyme substrate interactions and the function of lipid. *Journal of Biological Chemistry*, **251**, 5095–103.

Fujii, D.K. & Fulco, A.J. (1977). Biosynthesis of unsaturated fatty acids by bacilli-hyperinduction of desaturase synthesis. *Journal of Biological Chemistry*, **252**, 3660–70.

Fukushima, H., Martin, C.E., Iida, H., Kitajima, Y., Thompson, G.A. Jr. & Nozawa, Y. (1976). Changes in membrane lipid composition during temperature adaptation by a thermotolerant strain of *Tetrahymena pyriformis*. *Biochimica et Biophysica Acta*, **431**, 165–79.

Garda, H.A. & Brenner, R.R. (1984). Short-chain aliphatic alcohols increase rat liver microsomal membrane fluidity and affect the activites of some microsomal membrane-bound enzymes. *Biochimica et Biophysica Acta*, **769**, 160–70.

Garda, H.A. & Brenner, R.R. (1985). *In vitro* modification of cholesterol content of rat liver microsomes. Effects upon membrane 'fluidity' and activities of glucose-6-phosphatase and fatty acid desaturation systems. *Biochimica et Biophysica Acta*, **819**, 45–54.

Gil, H.A., Faust, J.R., Chin, D.J., Goldstein, J.L. & Brown, M.S. (1985). Membrane-bound domain of HMG-CoA reductase is required for sterol-enhanced degradation of the enzyme. *Cell*, **41**, 249–58.

Gombos, Z., Wada, H. & Murata, N. (1991). Direct evaluation of effects of fatty-acid unsaturation on the thermal properties of photosynthetic activities, as studied by mutation and transformation of Synechocystis PCC6803. *Plant Cell Physiology*, **32**, 205–11.

Gombos, Z., Wada, H. & Murata, N. (1992). Unsaturation of fatty acids in membrane lipids enhances tolerance of the cyanobacterium Synechocystis PCC6803 to low-temperature photoinhibition. *Proceedings of the National Academy of Sciences, USA*, **89**, 9959–63.

Gombos, Z., Wada, H. & Murata, N. (1994). The recovery of photosynthesis from low-temperature photoinhibition is accelerated by the unsaturation of membrane lipids: a mechanism of chilling tolerance. *Proceedings of the National Academy of Sciences, USA*, **91**, 8787–91.

Grayburn, W.S., Collins, G.B. & Hildebrand, D.F. (1992). Fatty acid alteration by a Delta 9 desaturase in transgenic tobacco tissue. *Biotechnology*, **10**, 675–8.

Hazel, J.R. & Landrey, S.R. (1988*a*). Time course of thermal adaptation in plasma membranes of trout kidney. II Molecular species composition. *American Journal of Physiology*, **255**, R628–34.

Hazel, J.R. & Landrey, S.R. (1988*b*). Time-course of thermal adaptation in plasma membranes of trout kidney. I Headgroup composition. *American Journal of Physiology*, **255**, R622–7.

Hazel, J.R. & Williams, E.E. (1990). The role of alterations in membrane lipid composition in enabling physiological adaptation of organisms to their physical environment. *Progress in Lipid Research*, **29**, 167–227.

Hazel, J.R., Williams, E.E., Livermore, R. & Mozingo, N. (1991). Thermal adaptation in biological membranes: functional significance of changes in phospholipid molecular species composition. *Lipids*, **26**, 277–82.

Horvath, I., Torok, Z., Vigh, L. & Kates, M. (1991). Lipid hydrogenation induces elevated 18:1-CoA desaturase activity in *Candida lipolytica* microsomes. *Biochimica et Biophysica Acta*, **1085**, 126–30.

Ishizaki, O., Nishida, I., Agata, K., Eguchi, G. & Murata, N. (1988). Cloning and nucleotide sequence of cDNA for the plastid glycerol-3-phosphate acyltransferase from squash. *FEBS Letters*, **238**, 424–30.

Jaworski, J.G. (1987) Biosynthesis of monoenoic and polyenoic fatty acids. In *Lipids: Structure and Function*. ed. Stumpf, P.K., vol. 9, pp. 159–174. Orlando, Florida: Academic Press.

Kaestner, K.H., Ntambi, J.M., Kelly, T.J. Jr. & Lane, M.D. (1989). Differentiation-induced gene expression in 3T3-L1 preadipocytes. A second differentially expressed gene encoding stearoyl-CoA desaturase. *Journal of Biological Chemistry*, **264**, 14755–61.

Kasai, R., Kitajima, Y., Martin, C.E., Nozawa, Y., Skriver, L. & Thompson, G.A. Jr. (1976). Molecular control of membrane properties during temperature acclimation. Membrane fluidity regulation of fatty acid desaturase action. *Biochemistry*, **15**, 5228–33.

Los, D., Horvath, I., Vigh, L. & Murata, N. (1993). The temperature-dependent expression of the desaturase gene *desA* in Synechocystis PCC6803. *FEBS Letters*, **318**, 57–60.

Macartney, A.I., Maresca, B. & Cossins, A.R. (1994). Acyl-CoA desaturases and the adaptive regulation of membrane lipid composition. In *Temperature Adaptation of Biological Membranes*, ed. A.R. Cossins, pp. 129–139. London: Portland Press.

Macartney, A.I., Tiku, P. & Cossins, A.R. (1995). An isothermal induction of the Δ^9-desaturase in carp hepatocytes. *Biochemica et Biophysica Acta* (submitted).

McKeon, T.A. & Stumpf, P.K. (1982). Purification and characterization of the stearoyl–acyl carrier protein desaturase and the acyl–acyl carrier protein thioesterase from maturing seeds of safflower. *Journal of Biological Chemistry*, **257**, 12141–7.

Maresca, B. & Cossins, A.R. (1994). Fatty feedback and fluidity. *Nature*, **365**, 606–7.

Martin, C.E., Hiramitsu, K., Kitajima, Y., Nozawa, Y., Skriver, L. & Thompson, G.A. Jr. (1976). Molecular control of membrane properties during temperature acclimation. Fatty acid desaturase regulation of membrane fluidity in acclimating *Tetrahymena* cells. *Biochemistry*, **15**, 5218–27.

Mihara, K. (1990). Structure and regulation of rat liver microsomal stearoyl-CoA desaturase gene. *Journal of Biochemistry, Tokyo*, **108**, 1022–9.

Morimoto, R., Tissieres, A. & Georgopoulos, C. (1994). *The Biology of Heat Shock Proteins and Molecular Chaperones*. pp. 1739–1740. New York: Cold Spring Harbor Press.

Murata, N., Ishizaki-Nishizawa, O., Higashi, S., Hayashi, H., Tasaki, Y. & Nishida, I. (1992). Genetically engineered alteration in the chilling sensitivity of plants. *Nature*, **356**, 710–3.

Nishida, I., Tasaka, Y., Shiraishi, H. & Murata, N. (1993). The gene and the RNA for the precursor to the plastid-located glycerol-3-phosphate acyltransferase of *Arabidopsis thaliana*. *Plant Molecular Biology*, **21**, 267–77.

Nozawa, Y., Iida, H., Fukushima, H., Ohki, K. & Ohnishi, S. (1974). Studies on *Tetrahymena* membranes. Temperature-induced alterations in fatty acid composition of various membrane fractions in *Tetrahymena pyriformis* and its effect on membrane fluidity as inferred by spin-label study. *Biochimica et Biophysica Acta*, **367**, 134–47.

Ntambi, J.M., Buhrow, S.A., Kaestner, K.H., Christy, R.J., Sibley, E., Kelly, T.J. Jr. & Lane, M.D. (1988). Differentiation-induced gene expression in 3T3-L1 preadipocytes. Characterization of a differentially expressed gene encoding stearoyl-CoA desaturase. *Journal of Biological Chemistry*, **263**, 17291–300.

Sakamoto, T., Los, D.A., Higashi, S., Wada, H., Nishida, I., Ohmori, M. & Murata, N. (1994). Cloning of omega 3 desaturase from cyanobacteria and its use in altering the degree of membrane-lipid unsaturation. *Plant Molecular Biology*, **26**, 249–63.

Sato, N. & Murata, N. (1981). Studies on the temperature shift-induced desaturation of fatty-acids in monogalactosyl diacylglycerol in the blue-green alga (Cyanobacterium), *Anabaena variabilis. Plant Cell Physiology*, **22**, 1043–50.

Schünke, M. & Wodtke, E. (1983). Cold-induced increase of Δ^9- and Δ^6-desaturase activities in endoplasmic reticulum of carp liver. *Biochimica et Biophysica Acta*, **734**, 70–5.

Sinesky, M. (1974). Homeoviscous adaptation–a homeostatic process that regulates the viscosity of membrane lipids in *E. coli. Proceedings of the National Academy of Sciences, USA*, **71**, 522–.

Strittmatter, P., Spatz, L., Corcoran, D., Rogers, M.J., Setlow, B. & Redline, R. (1974). Purification and properties of rat liver microsomal stearyl coenzyme A desaturase. *Proceedings of the National Academy of Sciences, USA*, **71**, 4565–9.

Strittmatter, P., Thiede, M.A., Hackett, C.S. & Ozols, J. (1988). Bacterial synthesis of active stearoyl-CoA desaturase lacking the 26-residue amino-terminal amino-acid sequence. *Journal of Biological Chemistry*, **263**, 2532–5.

Stubbs, C.D., Kouyama, T., Kinosita, K. & Ikegama, A. (1981). Effects of double bonds on the dynamic properties of the hydrocarbon region of lecithin bilayers. *Biochemistry*, **20**, 2800–10.

Stukey, J.E., McDonough, V.M. & Martin, C.E. (1989). Isolation and characterization of *OLE1*, a gene affecting fatty acid desaturation from *Saccharomyces cerevisiae. Journal of Biological Chemistry*, **264**, 16537–44.

Stukey, J.E., McDonough, V.M. & Martin, C.E. (1990). The *OLE1* gene of *Saccharomyces cerevisiae* encodes the delta 9 fatty acid desaturase and can be functionally replaced by the rat stearoyl-CoA desaturase gene. *Journal of Biological Chemistry*, **265**, 20144–9.

Swick, A.G. & Lane, D. (1992). Identification of a transcriptional repressor down-regulated during preadipocyte differentiation. *Proceedings of the National Academy of Sciences, USA*, **89**, 7895–9.

Thiede, M.A., Ozols, J. & Strittmatter, P. (1986). Construction and sequence of cDNA for rat liver stearoyl coenzyme A desaturase. *Journal of Biological Chemistry*, **261**, 13230–5.

Thiede, M.A. & Strittmatter, P. (1985). The induction and characterization of rat liver stearyl-CoA desaturase mRNA. *Journal of Biological Chemistry*, **260**, 14459–63.

Thompson, G.A. Jr. & Nozawa, Y. (1977). *Tetrahymena:* a system for studying dynamic alterations within the eukaryotic cell. *Biochimica et Biophysica Acta*, **472**, 55–92.

Tiku, P.E., Gracey, A.Y., Macartney, A.I., Beynon, R.B. & Cossins, A.R. (1996). Cold-induction of Δ^9-desaturase by transcriptional and post-translational mechanisms. *Science* (in press).

Vigh, L., Los, D.A., Horvath, I. & Murata, N. (1993). The primary signal in the biological perception of temperature: Pd-catalysed hydrogenation of membrane lipids stimulated the expression of the *desA* in *Synechocystis* PCC6803. *Proceedings of the National Academy of Sciences, USA*, **90**, 9090–4.

Wada, H., Gombos, Z. & Murata, N. (1990). Enhancement of chilling tolerance of a cyanobacterium by genetic manipulation of fatty acid desaturation. *Nature*, **347**, 200–3.

Wada, H., Schmidt, H., Heinz, E. & Murata, N. (1993). *In vitro* ferredoxin-dependent desaturation of fatty acids in cyanobacterial thylakoid membranes. *Journal of Bacteriology*, **175**, 544–7.

Wodtke, E. (1986). Adaptation of biological membranes to temperature: Modifications and their mechanisms in the eurythmic carp. *Biona Reports (Gustay Fischer, Stuttgart)*, **4**, 129–38.

Wodtke, E. & Cossins, A.R. (1991). Rapid cold-induced changes of membrane order and delta 9-desaturase activity in endoplasmic reticulum of carp liver: a time-course study of thermal acclimation. *Biochimica et Biophysica Acta*, **1064**, 343–50.

Wodtke, E., Teichert, T. & Konig, A. (1986). Control of membrane fluidity in carp upon cold stress: studies on fatty acid desaturases. In *Living in the Cold: Physiological and Biochemical Adaptations*, ed. H.C. Heller, pp. 35–42. Elsevier.

C.P. CUTLER, I.L. SANDERS, G. LUKE, N. HAZON and G. CRAMB

Ion transport in teleosts: identification and expression of ion transporting proteins in branchial and intestinal epithelia of the European eel

Introduction

Different environmental salinities are known to exert profound effects on the expression and activity of specific proteins associated with ion transport in migratory or estuarine aquatic vetebrates. The osmoregulatory adaptation of the ion transporting capacity is essential for the survival of teleost fish which inhabit both freshwater (FW) and seawater (SW) environments and in particular for catadromous and anadromous species which migrate between the two environments during their life cycle. The principles of osmoregulation and adaptation of teleosts to environments of differing salinity are generally accepted to involve concerted responses controlling the rate of drinking and the subsequent regulation and sometimes reversal of ion transport in the secretory/absorptive epithelia of the gill, gut, kidney and bladder. These processes ensure that the osmolality of the plasma varies little although there can be several orders of magnitude changes in the salinity of the environment.

When teleost fish, such as the European eel (*Anguilla anguilla*), are in an hypo-osmotic medium such as FW, they continually gain water across permeable body surfaces, primarily the gills, and this influx is balanced by the production of large quantities of dilute urine by the renal system. Salt losses from the body are reduced by the low body surface permeability and the absorption of ions from the food in the intestine and an efficient reabsorption of ions from the tubular fluid and urine in the kidney and bladder, respectively. Furthermore, some ions, and in particular calcium, are actively absorbed across the gills. When, as part of their natural life cycle, eels enter SW they encounter a change in the environmental osmolality of around 1000 mOsmol/kg. In order to prevent dehydration, an immediate drinking reflex occurs in response to elevation of the chloride ion concentration in the buccal cavity (Hirano, 1974). In the longer term, this increase in drinking is sustained by increases in plasma angiotensin II concentrations (Perott *et al.*, 1992; Tierney *et al.*, 1995) mediated by NaCl absorption and the dehydration associated with the loss of water across permeable body surfaces (Maetz & Skadhauge, 1968; Kirsch & Mayer-Gostan, 1973). The drinking response, which is essential for the maintenance of body fluid volume, results in the uptake of hyper-osmotic SW into the gut. The mechanism of water transport across the teleost intestine has not been fully elucidated; however, passive water absorption requires a reduction in the osmolality of the luminal fluid to below that of the body fluids. This process is thought to occur via desalination of the luminal contents by the active uptake of ions, initially in the oesophagus (Hirano & Mayer-Gostan, 1976; Nagashima & Ando, 1994) and then further, and in parallel with the uptake of water, in the intestine (Skadhauge, 1974; Hirano & Mayer-Gostan, 1976; Ando, Saski & Huang, 1986). Water conservation is enhanced by reduced urine production, and the sodium and chloride ions absorbed with the water in the gut are actively excreted from the body by the specialized epithelial 'chloride cells' located in the gill, opercular membrane and skin (Foskett *et al.*, 1983).

Chloride cells of the gill and operculum

Morphology

Teleost chloride cells have been extensively studied since they were originally identified in the 1930s (Keys & Willmer, 1932). Chloride cells in general have several characteristic morphological features which represent adaptations associated with the ion transporting function of the cell. Current evidence suggests that two types of chloride cells (α and β) exist in the FW-adapted fish gill, whereas only one type of chloride cell (α) is present along with the smaller accessory cells in the SW fish gill (Pisam *et al.*, 1993; see also Pisam & Rambourg, 1991 for references). The exception to this is the FW-adapted rainbow trout (*Salmo gairdneri*) where accessory cells may also be present (Pisam, Prunet & Rambourg, 1989). In water of low ion content, such as FW, gill chloride cells in trout species can be located in both the primary and secondary lamellae and have either flat or rounded apical surfaces (Laurent & Dunel, 1980; Laurent, 1984; Avella *et al.*, 1987; Pisam *et al.*, 1990). Chloride cells in other species such as the mullet or the eel are mainly located in the primary lamellae and have broad, shallow apical crypts sometimes with cellular mass protruding above the surface of the epithelium (Hossler, Ruby & McIlwain, 1979; Sardet, Pisam & Maetz, 1979; Laurent & Dunel, 1980; Laurent, 1984). Chloride cells in FW-adapted fish have an apically located vesiculo-tubular system and a tubular system which forms a broad, loose network, which is interspersed with mitochondria, and is continuous with the plasma membrane of the basolateral cell surface (see Pisam & Rambourg, 1991 for references). Deep multi-stranded high resistance tight junctions also occur between chloride cells and adjacent pavement/respiratory cells (Sardet *et al.*, 1979; Laurent & Dunel, 1980; Karnaky, 1986; Pisam *et al.*, 1990). In ion-poor soft water, the gills are known to be involved in the process of absorption of both sodium and chloride ions (Evans, 1975, 1980; Girard & Payan, 1977, 1980; Gairdaire *et al.*, 1985) and chloride cells have been suggested as a possible site of net influx of these ions (see Laurent *et al.*, 1994 for references).

When euryhaline teleosts migrate into SW, chloride cells undergo several morphological changes. Firstly β chloride cells degenerate and disappear over a period of days (Pisam, Caroff & Rambourg, 1987; Wendelaar Bonga & van der Meij, 1989) whereas the α chloride cells of the gill and the opercular membrane increase in both size and number, and smaller juxtaposed accessory cells appear (Yoshikawa *et al.*, 1993; see also Foskett *et al.*, 1983 and Pisam & Rambourg, 1991 for references). The function of the accessory cells has not yet been established; however, their appearance in SW-adapted animals suggests a role in the secretion of sodium and chloride ions.

Following SW adaptation, chloride cells generally have a more extensive tubular network, vesiculo-tubular system and an increased number of mitochondria. Tight junctions between adjacent chloride or accessory cells become shallower, and apical crypts deepen and narrow. The cytoplasmic processes of adjacent chloride and accessory cells become interdigitated especially in the region of the apical crypt (Karnaky, 1986; Pisam & Rambourg, 1991).

The ion transport model

Many studies have shown that the chloride cells in the gill and opercular epithelia of SW-adapted teleosts are the major sites of the active sodium and chloride secretion, allowing plasma ion homeostasis in this medium (Marshall & Nishioka, 1980; Foskett & Scheffey, 1982; Foskett *et al.*, 1983; Karnaky *et al.*, 1984). Ion substitution experiments have provided evidence that serosal sodium (Degnan, 1984) and potassium (Degnan & Zadunaisky, 1980*b*) ions are required for chloride secretion, and this suggests that the cellular entry of chloride ions may be mediated by a furosemide- or bumetanide-sensitive Na/K/2Cl cotransporter located in the basolateral membranes of the chloride cells (Degnan, Karnaky & Zadunaisky, 1977; Karnaky, Degnan & Zadunaisky, 1977; Zadunaisky, 1984; Davis & Shuttleworth, 1985; Karnaky, 1986; Zadunaisky *et al.*, 1995). The uptake and accumulation of chloride ions in the cell would then be driven by the favourable sodium gradient provided by the Na, K-ATPase (Silva *et al.*, 1977; Karnaky, 1986). In the model of ion transport first proposed by Silva *et al.* (1977; see Fig. 1), intracellular chloride ions would exit, down the electrical gradient created, through channels in the apical crypt, into the external medium. In addition to this model, chloride uptake is further aided by the recycling of potassium ions to the basolateral side through barium-sensitive channels (Degnan, 1985). Some potassium ions may also diffuse laterally and be secreted paracellularly in a similar fashion to the sodium ions (Marshall, 1981*b*). A role in active ion secretion has also been suggested for the polyanionic polysaccharide material located within the apical crypt and tubulo-vesicular system, but the part that

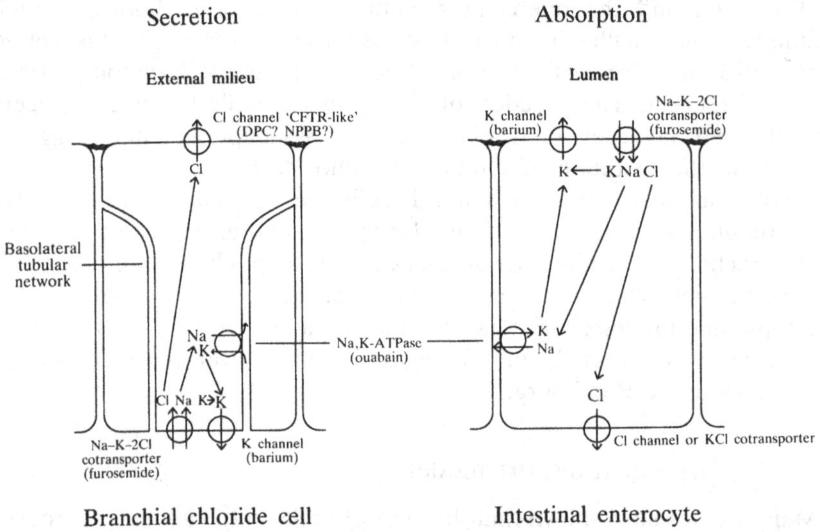

Fig. 1. A schematic diagram representing a basic working model of ion transport in the secretory chloride cells and absorptive enterocytes of the branchial and intestinal epithelium respectively. Items in parentheses represent known or possible (?) inhibitors of the transport proteins shown.

this material plays is far from clear (Karnaky & Philpott, 1969; Pisam, Sardet & Maetz, 1980; Bradley, 1981; Pisam *et al.*, 1983).

The Na,K-ATPase

Initial reports implicating the chloride cell as the main site of sodium and chloride secretion were based on the fact that high levels of the active sodium-transporting enzyme, Na,K-ATPase, were also localized in these cells (Hootman & Philpott, 1979; Ernst, Dodson & Karnaky; 1980; Karnaky, 1986; McCormick, 1990*a*; see also Foskett *et al.*, 1983 for additional references). Measurements of the specific enzymatic activity of the Na,K-ATPase in these preparations indicated that the chloride cells show the highest level of enzyme density measured in any epithelia (Karnaky *et al.*, 1976; Karnaky, 1986). Furthermore, the levels of Na,K-ATPase activity and abundance in gill and opercular epithelia have been shown to increase following adaptation of fish to hyper-osmotic media such as SW (Kamiya & Utida, 1969; Pfeiler & Kirschner, 1972; Towle, Gilman & Hempel, 1972; Sargent & Thomson, 1974; Sargent, Thomson & Bornancin, 1975; Ho & Chan, 1980; Madsen & Naarmanse, 1989; Mayer-Gostan & Lemaire, 1991; Pagliarani *et al.*, 1991; Almendras, Prunet & Boeuf, 1993; Nonnotte & Boeuf, 1995; see also Foskett *et al.*, 1983 for additional references). The Na,K-ATPase enzyme has been localized within the chloride cell to the basolateral tubular network (Karnaky *et al.*, 1976; Hootman & Philpott, 1979; Karnaky, 1986), suggesting that the Na,K-ATPase could only play an indirect role in active sodium transport by the accumulation of sodium ions in the tubular system. Sodium ions would then move into the external medium, by diffusing to the lateral extracellular spaces and then exiting paracellularly via 'leaky' tight junctions (Karnaky, 1986). This model of sodium transport has been supported by evidence from electrophysiological experiments examining the response of sodium fluxes to voltage clamping which suggests that sodium ion movement occurs only passively (Kirschner, Greenwald & Sanders, 1974; Degnan *et al.*, 1977; Degnan & Zadunaisky, 1979, 1980*a*). Sodium ions are secreted against their concentration gradient as a result of the electrochemical potential generated by the active transport of chloride ions (Zadunaisky, 1984; Karnaky, 1986). The involvement of Na,K-ATPase in the active secretion of chloride by chloride cells has also been demonstrated. Ouabain (a specific Na,K-ATPase inhibitor) reduced net chloride flux in chloride cells of the skin of the goby (*Gillichthys mirabilis*) (Marshall, 1981*a*) and the gills of the eel (*Anguilla rostrata*) (Silva *et al.*, 1977).

Chloride channels

Very little is known about the nature of the apical chloride conducting channels involved in chloride secretion. However, recent work on opercular chloride cells has demonstrated the presence of a channel which was inhibited by submillimolar concentrations of the chloride channel inhibitor diphenylamine-2-carboxylate (DPC; Anderson *et al.*, 1992) but only when applied to the basolateral but not apical surface (Marshall *et al.*, 1995; Zadunaisky *et al.*, 1995). Much higher concentrations (millimolar) of DPC or 5-nitro-2(3-phenylpropylamino)-benzoic acid (NPPB; another chloride channel inhibitor; Anderson *et al.*, 1992) were required to induce inhibition from the apical surface (Marshall *et al.*, 1995); however, apical application of the chloride channel (and Cl/HCO_3 exchange) blocker diisothiocyanostilbene disulphonate (DIDS) was without effect (Anderson *et al.*, 1992). Together, these results suggest the presence of a Cl channel similar in nature to the Cystic Fibrosis Transmembrane conductance Regulator (CFTR) (Riordan *et al.*, 1989; Anderson *et al.*, 1992); however, the concentration of DPC required to act on each membrane domain currently suggests that this channel is basolaterally rather than apically located. Chloride ion flux can also be inhibited by the mucosal application of Cu ions, although the mechanism of inhibition is unknown (Degnan, 1985).

Ion exchangers/cotransporters

Some reports have suggested that basolaterally located Na/H and Cl/HCO_3 exchangers are present in opercular chloride cells (Karnaky *et al.*, 1977; Zadunaisky *et al.*, 1995). It has been proposed that the rapid adaptation of the gill epithelium to high salinities is independent of hormonal changes, and occurs due to increased plasma osmolality. The increase in plasma osmolality causes chloride cell shrinkage and subsequent activation of bumetanide-sensitive Na/K/2Cl cotransporters and amiloride-sensitive Na/H exchangers, with a resultant increase in chloride efflux (Zadunaisky *et al.*, 1995). Ion substitution experiments on opercular epithelia have also demonstrated that, contrary to what might be expected, reductions in sodium or chloride ion concentrations in the external (mucosal) medium reduce chloride efflux, suggesting an additional regulatory role for the environmental salinity (Degnan & Zadunaisky, 1980*b*; Degnan, 1984).

Hormonal regulation of ion transport

A number of hormones have been implicated in the regulation of chloride cell proliferation and differentiation as well as the control of ion transport. In FW-adapted teleosts, prolactin is the main osmoregulatory hormone producing reductions in passive ion and water permeabilities as well as active and passive ion secretion. This probably occurs due to the differentiation of the β chloride cells and modification or de-differentiation of α chloride cells (Ogawa, 1974, 1975; Gallis, Lasserre & Belloc, 1979; Foskett, Machen & Bern, 1982*b*; Foskett *et al.*, 1983; Zadunaisky, 1984; Evans, 1990; Pisam *et al.*, 1993). Chloride cells have also been identified as a site of calcium absorption in the gills of FW teleosts (Payan, Mayer-Gostan & Pang, 1981; Ishihara & Mugiya, 1987; Flik & Verbost, 1993), and calcium uptake may be regulated by prolactin which, in conjunction with the hormones somatolactin and cortisol, exerts an hyper-calcaemic action (Flik & Verbost, 1993). A novel hypo-calcaemic hormone, stanniocalcin (Hypocalcin), which is produced and released by the corpuscles of Stannius, antagonizes the action of prolactin and inhibits gill calcium uptake (Flik & Verbost, 1993).

When fish are adapted to ion-poor FW, such as soft water, the hormone cortisol may also play a more active osmoregulatory role inducing the hypertrophy and differentiation of the chloride cells (see Laurent *et al.*, 1994 for references). During the course of, or following, SW adaptation, cortisol has been shown to increase both Na, K-ATPase enzyme abundance and activity as well as the size and density of chloride cells (Pickford *et al.*, 1970; Butler & Carmichael, 1972; Scheer & Langford, 1976; Hebgab & Hanke, 1984; Dange, 1986; Richman & Zaugg, 1987; McCormick & Bern, 1989; Madsen, 1990*a,b,c*; McCormick, 1990*b*; Bindon *et al.*, 1994; Foskett *et al.*, 1983). Growth hormone has also been shown to increase both Na, K-ATPase enzyme abundance and activity (Almendras *et al.*, 1993; Boeuf, 1993; Prunet *et al.*, 1994; Nonnotte & Boeuf, 1995; Seddiki *et al.*, 1995; Sakamoto, McCormick & Hirano, 1993) as well as the size and density of chloride cells (Prunet *et al.*, 1994; Sakamoto *et al.*, 1993) in salmonid species either during smoltification or after SW adaptation.

Several other hormones have been suggested as regulators of ion transport in teleost gill and opercular epithelia (for reviews see Foskett *et al.*, 1983; Evans, 1990). Ion transport can be stimulated by β_1-adrenergic agonists (Degnan & Zadunaisky, 1979; Marshall & Bern, 1980; Mendelson, Cherksey & Degnan, 1981; May *et al.*, 1984; Davis & Shuttleworth, 1985; Degnan, 1986; Marshall, Bryson & Garg, 1993), urotensin I (UI) (Marshall & Bern, 1981), glucagon and vasoactive intestinal peptide (VIP) (Foskett *et al.*, 1982*b*; Davis & Shuttleworth, 1985), and atrial natriuretic peptide (ANP) (Scheide & Zadunaiskym 1988) and can be inhibited by α_2-adrenergic agonists (Degnan *et al.*,

1977; Degnan & Zadunaisky, 1979; Mendelson *et al.*, 1981; Foskett *et al.*, 1982*a*; May *et al.*, 1984; Marshall *et al.*, 1993), cholinergic agonists (May & Degnan, 1985), somatostatin (Foskett & Hubbard, 1981), urotensin II (UII) (Marshall & Bern, 1979, 1981) and certain prostaglandins (Van Praag *et al.*, 1987). The mechanisms of stimulation of ion transport by β_1-adrenergic agonists, UI, glucagon and VIP have all been shown to involve elevation of intracellular cAMP levels (Mendelson *et al.*, 1981) and actions of these hormones have been mimicked by the direct application of this cyclic nucleotide or its hydrophobic analogues (Davis & Shuttleworth, 1985), the adenylate cyclase activator forskolin (May & Degnan, 1984; Davis & Shuttleworth, 1985; May & Degnan, 1985) and by phosphodiesterase inhibitors such as isobutylmethylxanthine (IBMX) (Marshall & Bern, 1981; Foskett *et al.*, 1982*b*; May & Degnan, 1984, 1985) or theophylline (Degnan *et al.*, 1977). The actions of cAMP may be mediated by the activation of protein kinase A (PKA) and the subsequent phosphorylation of either basolaterally located Na/K/2Cl cotransporters or apical located chloride channels, as has been shown in other vertebrate groups (Gadsby & Nairn, 1994; Haas, 1994). The regulation of basolateral Na,K-ATPase by phosphorylation/ dephosphorylation mechanisms can also not be ruled out (Bertorello & Katz, 1993). Inhibition of chloride cell ion transport by α_2-adrenergic or cholinergic agonists may involve changes in the intracellular calcium concentration; however, artificial manipulation with various effectors of intracellular Ca has produced contradictory results (Mendelson *et al.*, 1981; May & Degnan, 1985; Marshall *et al.*, 1993).

Ion transport in the intestine and oesophagus

In response to migration from hypo-osmotic to hyper-osmotic environments, fish lose body weight and show increases in plasma ion concentrations (Oide & Utida, 1967; Maetz & Skadhauge, 1968). The adaptation to the hyper-osmotic medium is associated with an increased rate of drinking (Kirsch & Mayer-Gostan, 1973; Hirano, 1974), which is proportional to the ionic strength of the external medium (Shehadeh & Gordon, 1969).

Morphology

Several morphological changes in both the intestine and oesophagus are associated with SW adaptation. In FW, the oesophagus has regular longitudinal folds lined mostly with stratified epithelium, with few mitochondrial-rich columnar cells. Following transfer to SW these features are gradually replaced by extended irregular meandering folds covered with an epithelium composed of mitochondrial-rich, columnar cells (Yamamoto & Hirano, 1978). The lateral spaces between the columnar cells in both the oesophagus and intestinal epithelia are distended (Yamamoto & Hirano, 1978; Nonnotte, Nonnotte & Leray, 1986). These morphological changes are associated with changes in epithelial ion transport. The FW-adapted eel oesophagus is impermeable to sodium and chloride ions as well as water (Hirano & Mayer-Gostan, 1976). However, after SW adaptation, sodium and chloride ions (but not water) are absorbed both actively and passively across the oesophagus (Hirano & Mayer-Gostan, 1976; Parmelee & Renfro, 1983; Nagashima & Ando, 1994). The ionic concentration of the luminal fluid is further reduced in the stomach and then further in the intestine until it becomes iso-osmotic with blood plasma (Shehadeh & Gordon, 1969) and water can then be absorbed along with the sodium and chloride in the intestine (Skadhauge, 1974; Ando, 1975; Hirano & Mayer-Gostan, 1976). The maximum rate of ion transport across the intestine is correlated with the osmolality of the external medium to which fish are adapted (Skadhauge, 1969) with up to 98% of the sodium and chloride and 80% of the water being absorbed from the imbibed SW (Shehadeh & Gordon, 1969). The mucus layer which coats the mucosal surface of both the oesophagus and intestine also has a functional role in ion absorption (Simonneaux *et al.*, 1987*a*; Simonneaux, Humbert & Kirsch, 1987*b*). The mucus layer acts as a diffusion barrier, slowing the movement of ions into or out of the region immediately adjacent to the epithelial cell surface and thus active ion absorption generates ionic gradients within the layer of mucus (Simonneaux *et al.*, 1987*b*).

Ion transport

The Na,K-ATPase

The mechanisms of sodium, chloride and water absorption in the intestine are not fully understood, but some features have been determined from physiological experiments. One component of the ion transporting system known to be involved in active sodium absorption, is the Na,K-ATPase. The importance of intestinal Na,K-ATPase for ion and water transport has been demonstrated, as the specific inhibitor ouabain, when added to the serosal surface, completely abolishes the active component of ion transport (Ando & Kobayashi, 1978; Field *et al.*, 1978; MacKay & Janicki, 1979; Ando, 1981; Parmalee & Renfro, 1983; Ando, 1985; Simonneaux *et al.*, 1987*b*; Baldisserotto & Mimura,

1994; Nagashima & Ando, 1994). Cytochemical techniques have identified the presence of Na,K-ATPase on the lateral surfaces of the oesophageal ion-transporting cells (Simonneaux *et al.*, 1987*a*). Furthermore, the level of Na,K-ATPase activity in the intestine has been shown to increase following transfer of fish from FW to SW (Jampol & Epstein, 1970; Colin *et al.*, 1985). It has been suggested that Na,K-ATPase actively accumulates sodium ions in the lateral spaces between intestinal columnar cells. Chloride ions can then diffuse down their electrochemical gradient from the cell to these local regions of high sodium concentration, creating an osmotic gradient for water to diffuse from the lumen (Field *et al.*, 1978; Nonnotte *et al.*, 1986).

Ion cotransporters

Evidence from ion substitution experiments and the application of pharmacological agents has suggested that the mechanisms of ion transport in the gut is similar, but reversed, to that found in the gill chloride cells (Fig. 1). An electroneutral Na/K/2Cl cotransporter is located in the apical brushborder membranes of the eel intestine (Musch *et al.*, 1982; Ando, 1985; Trischitta *et al.*, 1992). This has been demonstrated since substitution experiments have shown that the transport of sodium, potassium or chloride ions is interdependent (Ando & Kobayashi, 1978; Ando, 1980, 1983, 1985; Musch *et al.*, 1982; Trischitta *et al.*, 1992). Furthermore, ion transport across the intestine is inhibited by the loop diuretics furosemide (Eveloff *et al.*, 1980; Musch *et al.*, 1982; Baldisserotto & Mimura, 1994), bumetanide (Musch *et al.*, 1982; Halm, Krasny & Frizzell, 1985*b*; Trischitta *et al.*, 1992) and piretanide (Lau, 1985). It is possible that multiple cotransporter isoforms may be expressed in the intestine as work carried out using the winter flounder (*Pseudopleuronectes americanus*) indicates that three distinct protein size fractions can be cross-linked to the radiolabelled bumetanide analogue BSTBA (Suvitayavat *et al.*, 1994). Water influx across the intestine is highly correlated to that of the chloride ion influx, indicating that the cotransporter plays a major role, and is a potential control point, in volume regulation (Ando, 1983). Other types of cotransporter proteins have been implicated in SW-adapted teleosts. The presence of a sodium-independent potassium and chloride-coupled transport in SW-adapted intestine has suggested the presence of a K/Cl cotransporter (Ando, 1983, 1985), which may be basolaterally located (Halm, Krazny & Frizzell, 1985*a*). The expression of a thiazide-sensitive Na/Cl cotransporter has also been demonstrated in flounder intestine (Halm *et al.*, 1985*b*; Gamba *et al.*, 1993), although no thiazide-sensitive Na/Cl cotransporter activity could be demonstrated in the eel (Trischitta *et al.*, 1992). A bumetanide-sensitive Na/K/2Cl cotransporter is also present in some species of FW-adapted fish including the eel; however, the magnitude of the bumetanide-inhibitable short circuit current was only 30% of that found in SW-adapted eels (Trischitta *et al.*, 1989, 1992). In contrast, no bumetanide-sensitive Na/K/2Cl cotransporter was found in the gut of FW tilapia (*Oreochromis mossambicus*) (Groot & Bakker, 1988).

Other ion channels and exchangers

Potassium ions entering the intestinal enterocytes via the Na,K-ATPase or Na/K/2Cl cotransporter are passed into the intestinal lumen via barium-sensitive channels (Halm *et al.*, 1985*a,b*; Simonneaux *et al.*, 1987*b*; Ando, 1990; Trischitta *et al.*, 1992; Baldisserotto & Mimura, 1994). The secretion of potassium ions at the mucosal surface creates a high local concentration, the dispersion of which is inhibited by the mucus layer. This further reduces the potassium gradient across the apical membrane and stimulates potassium-dependent cotransport activity (Simonneaux *et al.*, 1987*b*). The presence of both a Na/H and a Cl/HCO_3 exchanger in the basolateral membrane of enterocytes has been proposed, owing to inhibition of ion transport by the specific inhibitors amiloride and DIDS respectively (Ando & Subramanyan, 1990; Baldisserotto & Mimura, 1994). The presence of an apically located Cl/HCO_3 exchanger has also been suggested (Howard & Ahern, 1988; Ando & Subramanyam, 1990; Baldisserotto & Mimura, 1994); however, other authors have failed to show any inhibition of transport after the application of DIDS to the luminal surface (Trischitta *et al.*, 1992).

Hormonal regulation of ion transport

A number of hormones have been shown to stimulate (including catecholamines, cortisol, and UII) or inhibit (including prolactin, ANP, VIP, acetylcholine, serotonin and histamine) ion transport in the intestine (for review, see Collie & Hirano, 1987). Catecholamines such as adrenaline have been shown to enhance ion transport via an α_2 receptor following inhibition by serotonin and/or methacholine (Ando & Kondo, 1993; Ando & Omura, 1993). Cortisol has been demonstrated to increase Na,K-ATPase activity (Pickford *et al.*, 1970; Epstein, Cynamon & McKay, 1971) and may also increase catecholamine receptor and Na/K/Cl cotransporter expression (Ando & Hara, 1994). The hormone UII also stimulates mucosal cotransport activity (Loretz, Freel & Bern, 1983; Loretz, Howard & Siegel, 1985). The effect of prolactin in the intestine is not consistent between species but in general reduces sodium and chloride absorption and lowers water permeability (Utida *et al.*, 1972; Morley, Chadwick & El Tounsy, 1981; Mainoya, 1982). ANP is a potent inhibitor of both ion and water absorption in the intestine of SW-adapted teleosts and inhibits Na/K/2Cl cotransport presumably by elevation of intracellular cGMP concentrations (O'Grady *et al.*, 1985; Ando, Kondo & Takei, 1992). Acetylcholine, serotonin, and histamine also inhibit both ion and water absorption but to a lesser extent than ANP (Mori & Ando, 1991; Ando *et al.*, 1992). VIP inhibits both ion and water absorption in the intestine of FW-adapted tilapia, possibly by elevation of intracellular cAMP concentrations (Mainoya & Bern, 1984). However, the effect of cAMP on the intestine in other teleost species can occasionally be stimulatory (Collie & Hirano, 1987). The action of the cyclic nucleotide cGMP is much clearer, as it consistently inhibits ion transport in teleost intestine (Frizzell *et al.*, 1979; Field, Smith & Bolton, 1980; Halm *et al.*, 1985*b*; O'Grady *et al.*, 1985; Ando *et al.*, 1992).

Molecular biology of ion transport in teleosts

Until recently there has been very little information in the literature on the molecular biology of genes coding for the proteins involved in ion transport process in the gill and intestine; however, new information on three of the main proteins involved in ion transport in these tissues is now available.

The Na,K-ATPase

A considerable amount is known about the molecular biology of Na,K-ATPase (or sodium pump) in vertebrate species other than teleost fish (for review see Lingrel & Kuntzweiler, 1994). The enzyme consists of two subunits, α and β, which are both necessary for function. At least five different isoforms of the α subunit have been isolated, which are encoded by different genes designated α1, α2, α3, α4 and λLMM5 (Shamraj & Lingrel, 1994; Lucking, Jorgenson & Meng, 1994; Lingrel & Kuntzweiler, 1994). The situation with the β subunit is more complicated as the closely related enzyme H, K-ATPase also possesses a β subunit, which has been shown to be interchangeable with those of Na,K-ATPase *in vitro* (Fambrough *et al.*, 1994). The β subunit isoform family contains five members, three of which (known as β1, β2 and β3) are known to associate with the Na,K-ATPase α subunits *in vivo* (Lingrel & Kuntzweiler, 1994). In teleost fish, a Na,K-ATPase α subunit cDNA has been cloned from the white sucker (*Catostomus commersoni*) (Schonrock *et al.*, 1991) and a small fragment of genomic DNA coding for an unidentified Na,K-ATPase α isoform has also been cloned from the rainbow trout (*Oncorhynchus miskiss*) (Kisen *et al.*, 1994). In the eel, full-length cDNA sequences of both the Na,K-ATPase α1 and β2 isoforms have been cloned and the putative amino acid sequences derived (Cutler *et al.*, 1995*a,b*; see Fig. 2).

Evidence suggesting the existence of a Na,K-ATPase α3 isoform in catfish (*Ictalurus punctatus*) brain has been reported using mammalian isoform speficic antibodies (Pressley, 1992). Work carried out in our laboratories has identified cDNA fragments from two Na,K-ATPase α subunit isoforms (α1 and α3) and six Na,K-ATPase 'β subunit-like' isoforms (designated β1, β3, β179, β185, β185b, β233) which have been cloned by RT–PCR amplification from eel brain. However, no evidence of the existence of direct counterparts of the mammalian Na,K-ATPase α2 or β2 isoforms could be found in eel brain tissue. The α3 cDNA hybridizes with either or both of two mRNA species, 5.4 and 3.6 kb in size, depending on the tissue. Sequencing studies indicate that four of the β isoform cDNA fragments (β179, β185, β185b, β233) encode previously uncharacterized proteins. The sizes of mRNA transcripts hybridizing with β isoform probes are 2.35, 2.40, 3.75, 6.50, not detected, and 2.35 kb for the β1, β3, β179, β185, β185b, and β233 isoforms, respectively.

With the now considerably expanded number of cDNA fragments, initial studies to characterize the expression of Na,K-ATPase subunit isoforms in various eel tissues were conducted to see which, if any, of the isoforms are expressed in the branchial and intestinal epithelia. Northern blots demonstrated that the β1 isoform is ubiquitously expressed in all tissues, and although the α1 isoform was expressed in most tissues there was no measureable expression in either cardiac or skeletal muscle (Cutler *et al.*, 1995*a,b*). Unlike the situation in mammals where expression of the Na,K-ATPase α3 isoform is mainly located in the brain (Young & Lingrel, 1987), detectable levels of expression of this isoform were also found in eel intestine, gill, kidney, ovary and eye. However, as the size and pattern of expression of the smaller (3.6 kb) α3 band was similar to that found with Northern blots of the α1 subunit (3.5 kb), this may indicate that the signals obtained are due to cross-reaction with the more prevalent α1 isoform expressed in the gill, intestine and kidney. Expression of an eel homologue of the Na,K-ATPase β3 isoform was also found in brain. Of the novel Na,K-ATPase β isoforms, expression of β233 is potentially the most interesting with respect to osmoregulation, as signals were detected in the gill, intestine and kidney as well as the brain. Of the other β

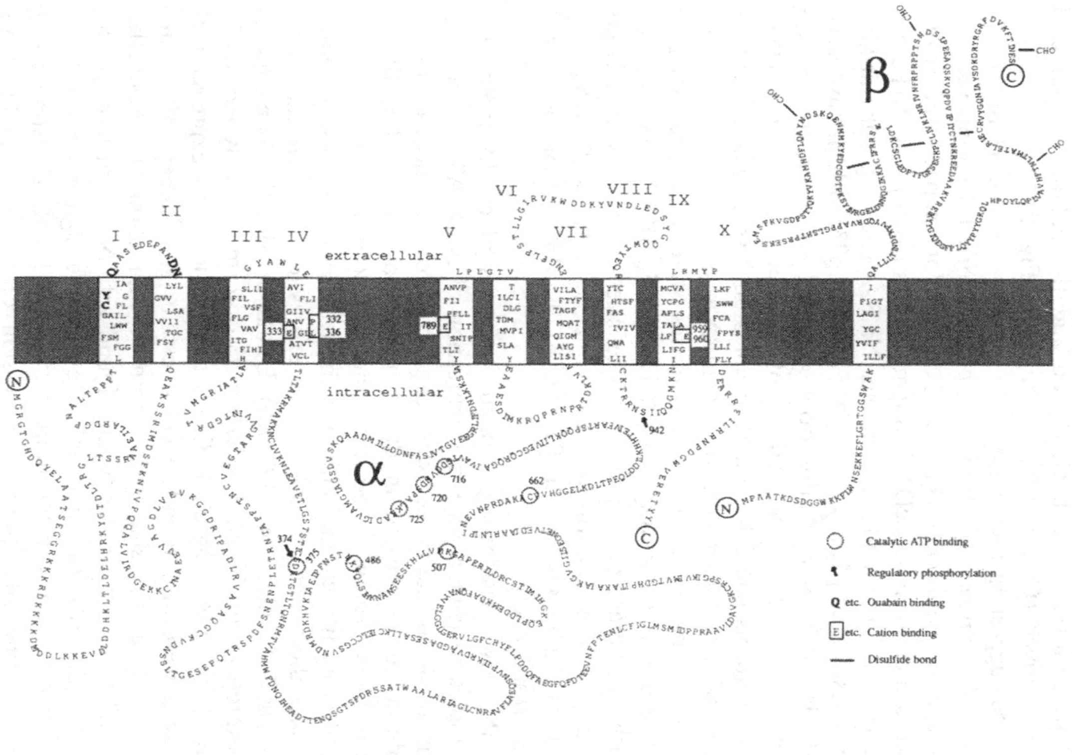

Fig. 2. Amino acid sequences and proposed membrane topographies of the eel Na,K-ATPase α1 and β1 subunits.

isoforms, β179 and β185 were found to be expressed only in brain, or brain and eye respectively. Expression of β185b has so far not been detected on any Northern blots using total RNA, suggesting that this isoform is expressed at very low levels in the tissues investigated.

In order to determine if there were any changes in expression of Na,K-ATPase α1 and β1 isoforms during adaptation of eels to different osmotic environments, a study was undertaken where fish were acutely transferred from FW to SW (Luke *et al.*, 1994). Changes in the activity and expression of Na,K-ATPase were associated with three phases. (*a*) Immediately upon transfer, Na,K-ATPase activity increased by a factor of three within 6 hours and then declined over the following 18 hours to activities found in the FW–FW control group. As there was no increase in α1 or β1 subunit mRNA production at this time, the mechanism of up-regulation of Na,K-ATPase activity is possibly associated with either the recruitment of pre-formed sequestered units to the plasma membrane or may be the result of a direct activation by phosphorylation/dephosphorylation mechanisms. This increase in activity may be related to the acutely elevated drinking response of eels upon entering SW (Hirano, 1974). (*b*) The initial peak of Na,K-ATPase activity was followed, some 3 days post-transfer, by a gradual and more prolonged increase in Na,K-ATPase activity which paralleled an increase in the abundance of α1 and β1 subunit mRNAs up to a maximum after 3 weeks of adaptation. Final activity levels were similar to the 6 hour peak and α1 and β1 subunit mRNAs increased by 3-fold and 6-fold, respectively. (*c*) Finally, in chronically adapted fish held in SW for 6 months, the increased levels of enzyme activity and α and β subunit mRNA expression were not sustained and by this time levels were reduced to near, but still above, those found in the FW–FW transferred group. These data suggest that the regulation of Na,K-ATPase in the branchial epithelium is more complex than previously recognized. The reasons for the reductions in activity are unclear, but may reflect changes in expression of other branchial epithelial transporting systems, or may result from a reduction in total body water loss and therefore a reduced requirement for drinking.

The optimum time point to study changes in Na,K-ATPase mRNA expression was approximately 3 weeks after SW transfer. A further study investigating expression 3 weeks post-transfer to various salinities was undertaken to identify changes in the mRNA abundance of the four Na,K-ATPase isoforms expression in the gill (α1, α3, β1 and β233). As found in the previous study, the levels of Na,K-ATPase α1 and β1 mRNAs increased after transfer of eels from FW to SW. Furthermore, even higher levels of α1 and β1 mRNAs were detected in the gills of fish adapted to artificially prepared 200% SW. The levels of Na,K-ATPase α1 and β1 mRNAs found in iso-osmotic saline were approximately the same as those in FW-adapted fish (Cutler *et al.*, 1995*a,b*). The patterns of expression of the Na,K-ATPase α3 and β233 isoform mRNAs were similar in that increases in the expression of both isoforms were only found in eels adapted to 200% SW.

The Northern blots demonstrated that extremely high levels of expression of Na,K-ATPase α1, α3, β1, and β233 isoform mRNAs are also found in the intestine and the levels of expression of all four Na,K-ATPase isoform mRNAs are greatly increased in mid-gut samples taken from SW-adapted compared to FW-adapted eels.

The Na/K/2Cl-cotransporter

The cloning of two related cotransporters from dogfish rectal gland (a bumetanide/furosemide-sensitive Na/K/2Cl cotransporter; Xu *et al.*, 1994) and from winter flounder bladder (a thiazide-sensitive Na/Cl cotransporter; Gamba *et al.*, 1993) allowed a strategy to be developed for the cloning of members of the eel cotransporter gene family using RT–PCR amplification with brain, intestine and gill mRNAs. Using an eel gill mRNA sample, a cDNA fragment was isolated (designated cot 1) which was most similar to the 'secretory type' cotransporters cloned from the dogfish rectal gland (60% amino acid homology; Xu *et al.*, 1994) and mouse kidney epithelial cells (Delpire *et al.*, 1994). However, the homology of the derived cot 1 amino acid sequence to the dogfish or mouse cotransporter sequences is not sufficient to allow it to be positively identified as the eel homologue of these proteins. Despite being cloned from the gill, Northern blot analysis revealed that the major sites of expression of the 9.0 kb cot 1 mRNA were the intestine and kidney, although cot 1 may be expressed in other tissues at lower levels. A preliminary study has demonstrated that cot 1 may be important for the secretion of ions in the intestine in FW, as an increase in cot 1 mRNA expression was found in intestinal mid-gut samples from fish adapted to FW compared to their SW-adapted counterparts. These data suggest that the intestine may have the capacity to secrete ions into the lumen while the eel is in FW environments. This may be important for the generation of ion gradients essential for nutrient uptake; however, the reported electrical properties of the eel intestine are consistent with the occurrence of net ion absorption in FW-adapted animals (Trischitta *et al.*, 1992; Ando & Hara, 1994).

A second cotransporter cDNA fragment was cloned from eel intestine (designated cot 2); the derived amino acid sequence of this clone was most similar to the sequence of a second absorptive type isoform of the bumetanide-sensitive Na/K/2Cl cotransporter recently discovered in mammals (53% amino acid homology; Gamba *et al.*, 1994; Payne & Forbush III, 1994) although, unlike the mammalian isoform which is expressed exclusively in the kidney, the 3.8 kb cot 2 mRNA was expressed at high levels in both intestine and bladder, which indicates that cot 2 may not be a direct homologue of the mammalian isoform. In contrast to cot 1, the expression of cot 2 mRNA in the intestine was dramatically increased when eels were adapted to SW environments. The increase in expression of cot 2 mRNA suggests that the increase in bumetanide-sensitive ion absorption found after SW adaptation of eels (Trischitta *et al.*, 1992; Ando & Hara, 1994) may be due to changes in gene expression (Ando & Hara, 1994) rather than, as previously suggested, activation of existing cotransporters (Trischitta *et al.*, 1992).

More recently a further cDNA fragment (designated cot 4) has been cloned which shares highest homology with the thiazide-sensitive Na/Cl cotransporters cloned from winter flounder and rat (Gamba *et al.*, 1993, 1994). Reverse transcriptase PCR (RT-PCR) experiments suggest that this gene is expressed in the intestine of SW-adapted eels. This is in direct contrast to the report of Trischitta *et al.* (1992) who found no evidence for the presence of a hydrochlorothiazide-sensitive Na/Cl cotransporter in either SW- or FW-adapted eel intestine. No information is yet available as to whether cot 4 is expressed in the intestine of the FW-adapted eel.

Finally, a fourth cotransporter cDNA fragment (designated cot 3) was cloned from eel brain and low levels of expression of three mRNA species (3.3, 3.6 and 5.5 kb) were detected in Northern blots of brain total RNA. Furthermore, Northern blot signals for cot 3 have also been detected in gill poly A+ RNA. The derived amino acid sequence of cot 3 shares the highest level of homology with that of cot 1 suggesting that it is also a member of the dogfish rectal gland-like Na/K/2Cl cotransporter sub-family. The location of cot 3 expression to the brain suggests that this cDNA may represent a novel cotransporter isoform and a possible homologous counterpart of a unique cotransporter demonstrated to exist in brain capillary endothelial cells (Vigne, Lopez Farre & Frelin, 1994).

The cystic fibrosis transmembrane conductance regulator (CFTR)

The third type of protein which is implicated in branchial epithelial ion transport is the apical chloride channel which may be related to the mammalian CFTR chloride channel which, when defective, is associated with the disease cystic fibrosis (Riordan *et al.*, 1989). As the CFTR gene product was a possible candidate for the ion channel involved in active chloride transport in the eel gill, attempts were made to identify a 'CFTR-like' chloride channel using RT–PCR amplification of eel gill poly A^+ RNA. Using this approach, a cDNA fragment was obtained, the amino acid sequence of which shares only 60% homology with the CFTR sequences reported for other species, which is 15% lower than that found between the evolutionarily more distant dogfish and mammals (Marshall *et al.*, 1991). The low homology of the 'CFTR-like' cDNA with sequences from other species suggests that either this cDNA represents a second isoform of CFTR in the eel or that the gene is functionally different in eels and therefore may explain why there are such marked changes in the amino acid sequence and possibly the proteins structure. Despite the low level of overall amino acid homology, specific conserved sequences found in all the CFTR genes so far sequenced are also present in the eel 'CFTR-like' cDNA fragment indicating that the eel gene is an homologue of the CFTR gene family and not another ABC transporter such as members of the multi-drug-resistant (MDR) gene family (Riordan *et al.*, 1989). The level of expression of the 5.45 kb 'CFTR-like' chloride channel mRNA was very low and was only detectable on blots of gill poly A^+ RNA. The low level of expression of the eel CFTR homologue in gill tissue, suggests that its expression may be restricted to a small subset of cells in the branchial epithelium such as the chloride cells; however, the cellular location of the eel CFTR mRNA remains to be determined.

Ca-ATPase

During experiments to clone other Na,K-ATPase α isoforms, an eel cDNA was also isolated which was 76% homologous to a ubiquitously expressed rat Ca-ATPase cDNA sequence (Gunteski-Hamblin, Clarke & Shull, 1992), which itself was also homologous to the PMR (plasma membrane ATPase-related) family of Golgi apparatus Ca-ATPases from yeast (Rudolph *et al.*, 1989). Northern blot analysis of Ca-ATPase mRNA expression shows that the eel Ca-ATPase homologue has two mRNA species (5.45 and 3.85 kb) and is expressed in the gills and intestine, with small amounts also expressed in the kidney and spleen. Initial studies have indicated that eel Ca-ATPase mRNA expression may be increased in FW gill compared to SW gill tissues. However, the role which this gene might play in the absorption of Ca ions by the gills of FW eels remains unclear.

Although substantial advances have been made to identify and characterize some of the transport proteins involved in the absorption and secretion of ions in the gill and intestine, further work is required to determine the factors controlling the expression and activity of these proteins and to establish the identity and mechanisms of action of the hormonal and environmental factors which regulate these processes.

Acknowledgements

We would like to thank the Natural Environment Research Council (NERC) (Grant GR3/7379A) and the Wellcome Trust (Grant 035333/Z/92/Z) for supporting different aspects of this work.

References

Almendras, J.M.E., Prunet, P. & Boeuf, G. (1993). Responses of a non-migratory stock of brown trout, *Salmo trutta*, to ovine growth hormone treatment and seawater exposure. *Aquaculture*, **114**, 169–79.

Anderson, M.P., Sheppard, D.N., Berger, H.A. & Welsh, M.J. (1992). Chloride channels in the apical membrane of normal and cystic fibrosis airway and intestinal epithelia. *American Journal of Physiology*, **263**, L1–14.

Ando, M. (1975). Intestinal water transport and chloride pump in relation to sea-water adaptation of the eel *Anguilla japonica. Comparative Biochemistry and Physiology*, **52A**, 229–33.

Ando, M. (1980). Chloride-dependent sodium and water transport in the seawater eel intestine. *Journal of Comparative Physiology*, **138**, 87–91.

Ando, M. (1981). Effects of ouabain on chloride movements across the seawater eel intestine. *Journal of Comparative Physiology*, **145**, 73–9.

Ando, M. (1983). Potassium-dependent chloride and water transport across the seawater eel intestine. *Journal of Membrane Biology*, **73**, 125–30.

Ando, M. (1985). Relationship between coupled Na^+ -K^+ -Cl^- transport and water absorption across the seawater eel intestine. *Journal of Comparative Physiology*, **155**, 311–17.

Ando, M. (1990). Effects of bicarbonate on salt and water transport across the intestine of the seawater eel. *Journal of Experimental Biology*, **150**, 367–79.

Ando, M. & Hara, I. (1994). Alteration of sensitivity to various regulators in the intestine of the eel following seawater acclimation. *Comparative Biochemistry and Physiology*, **109A**, 447–53.

Ando, M. & Kobayashi, M. (1978). Effects of stripping of the outer layers of the eel intestine on salt and water transport. *Comparative Biochemistry and Physiology*, **61A**, 497–501.

Ando, M. & Kondo, K. (1993). Noradrenaline antagonizes effects of serotonin and acetylcholine in the seawater eel intestine. *Journal of Comparative Physiology B*, **163**, 59–63.

Ando, M., Kondo, K. & Takei, Y. (1992). Effects of eel atrial natriuretic peptide on NaCl and water transport across the intestine of the seawater eel. *Journal of Comparative Physiology B*, **162**, 436–9.

Ando, M. & Omura, E. (1993). Catecholamine receptor in the seawater eel intestine. *Journal of Comparative Physiology B*, **163**, 64–9.

Ando, M. & Sasaki, H. & Huang, K.C. (1986). A new technique for measuring water transport across the seawater eel intestine. *Journal of Experimental Biology*, **122**, 257–68.

Ando, M. & Subramanyam, M.V.V. (1990). Bicarbonate transport systems in the eel intestine of the seawater eel. *Journal of Experimental Biology*, **150**, 381–94.

Avella, M., Masoni, A., Bornancin, M. & Mayer-Gostan, N. (1987). Gill morphology and sodium influx in the rainbow trout (*Salmo gairdneri*) acclimated to artificial freshwater environments. *Journal of Experimental Zoology*, **241**, 159–69.

Baldisserotto, B. & Mimura, O.M. (1994). Ion transport across the isolated intestinal mucosa of *Anguilla anguilla* (Pisces). *Comparative Biochemistry and Physiology*, **108A**, 297–302.

Bertorello, A.M. & Katz, A.I. (1993). Short-term regulation of renal Na–K-ATPase activity: physiological relevance and cellular mechanisms. *American Journal of Physiology*, **265**, F743–55.

Bindon, S.D., Gilmour, K.M., Fenwick, J.C. & Perry, S.F. (1994). The effects of branchial chloride cell proliferation on respiratory function in the rainbow trout *Oncorhynchus mykiss*. *Journal of Experimental Biology*, **197**, 47–63.

Boeuf, G. (1993). Salmonid smolting: a pre-adaptation to the oceanic environment. In *Fish Ecophysiology*, ed. J.C. Rankin & F.B. Jensen, pp. 105–35. London: Chapman & Hall.

Bradley, T.J. (1981). Improved visualisation of apical vesicles in chloride cells of fish gills using an osmium quick-fix technique. *Journal of Experimental Zoology*, **217**, 185–98.

Butler, D.G. & Carmichael, F.J. (1972). (Na^+ -K^+)-ATPase activity in eel (*Anguilla rostrata*) gills in relation to changes in environmental salinity: role of adrenocortical steroids. *General and Comparative Endocrinology*, **19**, 421–7.

Colin, D.A., Nonnette, G., Leray, C. & Nonnotte, L. (1985). Na transport and enzyme activities in the intestine if the freshwater and sea-water adapted trout (*Salmo gairdnerii* R.). *Comparative Biochemistry and Physiology*, **81A**, 695–8.

Collie, N.L. & Hirano, T. (1987). Mechanisms of hormone actions on intestinal transport. In *Vetebrate Endocrinology: Fundamentals and Biomedical Implications. Volume 2. Regulation of Water and Electrolytes*, ed. P.K.T. Pang & M.P. Schreibman, pp. 239–270. London: Academic Press.

Cutler, C.P., Sanders, I.L., Hazon, N. & Cramb, G. (1995*a*). Primary sequence, tissue specificity and expression of the Na^+,K^+-ATPase α1 subunit in the European eel (*Anguilla anguilla*). *Comparative Biochemistry and Physiology*, **111B**, 567–73.

Cutler, C.P., Sanders, I.L., Hazon, N. & Cramb, G. (1995*b*). Primary sequence, tissue specificity and expression of the Na^+,K^+-ATPase β1 subunit in the European eel (*Anguilla anguilla*). *Fish Physiology and Biochemistry*, **14**, 423–9.

Dange, A.D. (1986). Branchial Na^+–K^+-ATPase activity in freshwater and saltwater acclimated tilapia. *Oreochromis (Sarotherodon) mossambicus*: effects of cortisol. *General and Comparative Endocrinology*, **62**, 341–3.

Davis, M.S. & Shuttleworth, T.J. (1985). Peptidergic and adrenergic regulation of electrogenic ion transport in isolated gills of the flounder (*Platichthys flesus* L.). *Journal of Comparative Physiology*, **155**, 471–8.

Degnan, K.J. (1984). The sodium and chloride dependence of chloride secretion by the opercular epithelium. *Journal of Experimental Zoology*, **231**, 11–17.

Degnan, K.J. (1985). The role of K^+ and Cl^- conductances in chloride secretion by the opercular epithelium. *Journal of Experimental Zoology*, **236**, 19–25.

Degnan, K.J. (1986). Cyclic AMP stimulation of Cl^- secretion by the opercular epithelium: the apical membrane chloride conductance. *Journal of Experimental Zoology*, **238**, 141–6.

Degnan, K.J., Karnaky, K.J.Jr. & Zadunaisky, J.A. (1977). Active chloride transport in the *in vitro* opercular skin of a teleost (*Fundulus heteroclitus*), a gill-like epithelium rich in chloride cells. *Journal of Physiology*, **271**, 155–91.

Degnan, K.J. & Zadunaisky, J.A. (1979). Open-circuit sodium and chloride fluxes across isolated opercular epithelia from the teleost *Fundulus heteroclitus*. *Journal of Physiology*, **294**, 483–95.

Degnan, K.J. & Zadunaisky, J.A. (1980*a*). Passive sodium movements across the opercular epithelium: the paracellular shunt pathway and ionic conductance. *Journal of Membrane Biology*, **55**, 175–85.

Degnan, K.J. & Zadunaisky, J.A. (1980*b*). Ionic contribution to the potential and current across the opercular epithelium. *American Journal of Physiology*, **238**, R231–9.

Delpire, E., Rauchman, M.I., Hebert, S.C. & Gullans, S.R. (1994). Molecular cloning and chromosome localization of a putative basolateral Na^+–K^+-$2Cl^-$ cotransporter from mouse inner medullary collecting duct (mIMCD-3) cells. *Journal of Biological Chemistry*, **269**, 25677–83.

Epstein, F.H., Cynamon, M. & McKay, W. (1971). Endocrine control of Na–K-ATPase and seawater adaptation in *Anguilla rostrata*. *General and Comparative Endocrinology*, **16**, 323–8.

Ernst, S.A., Dodson, W.C. & Karnaky, K.J.Jr. (1980). Structural diversity of occluding junctions in the low-resistance chloride-secreting opercular epithelium of seawater-adapted killifish (*Fundulus heteroclitus*). *Journal of Cell Biology*, **87**, 488–97.

Evans, D.H. (1975). Ionic exchance mechanisms in fish gills. *Comparative Biochemistry and Physiology*, **51A**, 491–5.

Evans, D.H. (1980). Kinetic studies of ion transport by fish gill epithelium. *American Journal of Physiology*, **238**, R224–30.

Evans, D.H. (1990). An emerging role for a cardiac peptide hormone in fish osmoregulation. *Annual Review of Physiology*, **52**, 43–60.

Eveloff, J., Field, M., Kinne, R. & Murer, H. (1980). Sodium-cotransport systems in intestine and kidney of the winter flounder. *Journal of Comparative Physiology*, **135**, 175–82.

Fambrough, D.M., Lemas, M.V., Hamrick, M., Emerick, M., Renaud, K.J., Inman, E.M., Hwang, B. & Takeyasu, K. (1994). Analysis of subunit assembly of the Na–K-ATPase. *American Journal of Physiology*, **266**, C579–89.

Field, M., Karnaky, K.J.Jr., Smith, P.L., Bolton, J.E. & Kinter, W.B. (1978). Ion transport across isolated mucosa of the winter flounder, *Pseudopleuronectes americanus*. I. Functional and structural properties of cellular and paracellular pathways for Na and Cl. *Journal of Membrane Biology*, **41**, 265–93.

Field, M., Smith, P.L. & Bolton, J.E. (1980). Ion transport across isolated mucosa of the winter flounder, *Pseudopleuronectes americanus*. II. Effects of cAMP. *Journal of Membrane Biology*, **55**, 157–63.

Flik, G. & Verbost, P.M. (1993). Calcium transport in fish fills and intestine. *Journal of Experimental Biology*, **184**, 17–29.

Foskett, J.K., Bern, H.A., Machen, T.E. & Conner, M. (1983). Chloride cells and the hormonal control of teleost osmoregulation. *Journal of Experimental Biology*, **106**, 255–81.

Foskett, J.K. & Hubbard, D.M. (1981). Hormonal control of chloride secretion by teleost opercular membrane. *Annals of the New York Academy of Science*, **372**, 643.

Foskett, J.K., Hubbard, D.M., Machen, T.E. & Bern, H.A. (1982*a*). Effects of epinephrine, glucagon and vasoactive intestinal peptide on chloride secretion by teleost opercular membrane. *Journal of Comparative Physiology*, **146**, 27–34.

Foskett, J.K., Machen, T.E. & Bern, H.A. (1982*b*). Chloride secretion and conductance of teleost opercular membrane: effects of prolactin. *American Journal of Physiology*, **242**, R380–9.

Foskett, J.K. & Scheffey, C. (1982). The chloride cell: definitive identification as the salt-secretory cell in teleosts. *Science*, **215**, 164–6.

Frizzell, R.A., Smith, P.L., Vosbugh, E. & Field, M. (1979). Coupled sodium-chloride influx across brush border of flounder intestine. *Journal of Membrane Biology*, **46**, 27–39.

Gadsby, D.C. & Nairn, A.C. (1994). Regulation of CFTR channel gating. *Trends in Biochemical Sciences*, **19**, 513–18.

Gairdaire, E., Avella, M., Isaia, J., Bornancin, M. & Mayer-Gostan, N. (1985). Estimation of sodium uptake through the gill of the rainbow trout *Salmo gairdneri. Experimental Biology*, **44**, 181–9.

Gallis, J-L., Lasserre, P. & Belloc, F. (1979). Freshwater adaptation in the euryhaline teleost, *Chelon labrosus. General and Comparative Endocrinology*, **38**, 1–10.

Gamba, G., Miyanoshita, A., Lombardi, M., Lytton, J., Lee, W.S., Hediger, M.A. & Hebert, S.C. (1994). Molecular cloning, primary structure, and characterisation of two members of the mammalian electroneutral sodium–(potassium)–chloride cotransporter family expressed in kidney. *Journal of Biological Chemistry*, **269**, 17713–22.

Gamba, G., Saltzberg, S.N., Lombardi, M., Miyanoshita, A., Lytton, J., Hediger, M.A., Brenner, B.M. & Hebert, S.C. (1993). Primary structure and functional expression of a cDNA encoding the thiazide-sensitive, electroneutral sodium-chloride cotransporter. *Proceedings of the National Academy of Sciences, USA*, **90**, 2749–53.

Girard, J.P. & Payan, P. (1977). Kinetic analysis of sodium and chloride influxes across the gills of the trout in fresh water. *Journal of Physiology*, **273**, 195–209.

Girard, J.P. & Payan, P. (1980). Ion exchanges through respiratory and chloride cells in freshwater- and seawater-adapted teleosteans. *American Journal of Physiology*, **238**, R260–8.

Groot, J.A. & Bakker, R. (1988). NaC1 transport in vertebrate intestine. In *Advances in Comparative and Environmental Physiology. 1. NaCl Transport in Epithelia*, ed. R. Greger, pp. 103–152. Berlin: Springer.

Gunteski-Hamblin, A.-M., Clarke, D.M. & Shull, G.E. (1992). Molecular cloning and tissue distribution of alternatively spliced mRNAs encoding possible mammalian homologues of the yeast secretory pathway calcium pump. *Biochemistry*, **31**, 7600–8.

Haas, M. (1994). The Na-K-Cl cotransporters. *American Journal of Physiology*, **267**, C869–85.

Halm, D.R., Krasny, E.J.Jr. & Frizzell, R.A. (1985*a*). Electrophysiology of flounder intestinal mucosa. I. Conductance properties of the cellular and paracellular pathways. *Journal of General Physiology*, **85**, 843–64.

Halm, D.R., Krasny, E.J.Jr. & Frizzell, R.A. (1985*b*). Electrophysiology of flounder intestinal mucosa. II. Relation of the electrical potential profile to coupled NaCl absorptions. *Journal of General Physiology*, **85**, 865–83.

Hebgab, S.A. & Hanke, W. (1984). The significance of cortisol for osmoregulation in carp (*Cyrinus carpio*) and tilapia (*Sarotherodon mossambicus*). *General and Comparative Endocrinology*, **54**, 409–17.

Hirano, T. (1974). Some factors regulating water intake by the eel *Anguilla japonica*. *Journal of Experimental Biology*, **61**, 737–47.

Hirano, T. & Mayer-Gostan, N. (1976). Eel esophagus as an osmoregulatory organ. *Proceedings of the National Academy of Sciences, USA*, **73**, 1348–50.

Ho, S.-M. & Chan, D.K.O. (1980). Branchial ATPases and ionic transport in the eel, *Anguilla japonica*-I. Na^+,K^+-ATPase. *Comparative Biochemistry and Physiology*, **66B**, 255–60.

Hootman, S.R. & Philpott, C.W. (1978). Rapid isolation of chloride cells from pinfish gill. *Anatomical Record*, **190**, 687–702.

Hootman, S.R. & Philpott, C.W. (1979). Ultracytochemical localisation of Na^+,K^+-activated ATPase in chloride cells from the gills of a euryhaline teleost. *Anatomical Record*, **193**, 99–130.

Hossler, F.E., Ruby, J.R. & McIlwain, T.D. (1979). The gill arch of the mullet, *Mugil cephalus*. II. Modification in surface ultrastructure and Na,K-ATPase content during adaptation to various salinities. *Journal of Experimental Zoology*, **208**, 399–406.

Howard, J.N. & Ahern, G.A. (1988). Parallel antiport mechanisms for Na^+ and Cl^- transport in herbivorous teleost intestine. *Journal of Experimental Biology*, **135**, 65–76.

Ishihara, A. & Mugiya, Y. (1987). Ultrastructural evidence of calcium uptake by chloride cells in the gills of goldfish, *Carassius auratus*. *Journal of Experimental Zoology*, **242**, 121–9.

Jampol, L.M. & Epstein, F.H. (1970). Sodium–potassium-activated adenosine triphosphatase and osmotic regulation by fishes. *American Journal of Physiology*, **218**, 607–11.

Kamiya, M. & Utida, S. (1969). Changes in activity of sodium–potassium-activated adenosinetriphosphatase in gills during adaptation of the Japanese eel to seawater. *Comparative Biochemistry and Physiology*, **26**, 675–85.

Karnaky, K.J.Jr. (1986). Structure and function of the chloride cell of *Fundulus heteroclitus* and other teleosts. *American Zoologist*, **26**, 209–24.

Karnaky, K.J.Jr., Degnan, K.J., Garretson, L.T. & Zadunaisky, J.A. (1984). Identification and quantification of mitochondria-rich cells in transporting epithelia. *American Journal of Physiology*, **246**, R770–5.

Karnaky, K.J.Jr., Degnan, K.J. & Zadunaisky, J.A. (1977). Chloride transport across isolated opercular epithelium of killifish: a membrane rich in chloride cells. *Science*, **195**, 203–5.

Karnaky, K.J.Jr., Kinter, L.B., Kinter, W.B. & Sterling, C.E. (1976). Teleost chloride cell. II. Autoradiographic localization of gill Na, K-ATPase in killifish *Fundulus heteroclitus* adapted to low and high salinity environments. *Journal of Cell Biology*, **70**, 157–77.

Karnaky, K.J.Jr. & Philpott, C.W. (1969). The cytochemical demonstration of surface-associated polyanions in a cell specialized for electrolyte transport. *Journal of Cell Biology*, **43**, 64a–5a.

Keys, A. & Willmer, E.N. (1932). 'Chloride secreting cells' in the gill of fishes, with special reference to the common eel. *Journal of Physiology*, **76**, 368–77.

Kirsch, R. & Mayer-Gostan, N. (1973). Kinetics of water and chloride exchanges during adaptation of the European eel to seawater. *Journal of Experimental Biology*, **58**, 105–21.

Kirschner, L.B., Greenwald, L. & Sanders, M. (1974). On the mechanism of sodium extrusion across the irrigated gill of seawater-adapted rainbow trout (*Salmo gairdneri*). *Journal of General Physiology*, **64**, 148–65.

Kisen, G., Gallis, C., Auperin, B., Klungland, H., Sandra, O., Prunet, P. & Anderson, O. (1994). Northern blot analysis of the Na^+,K^+-ATPase α-subunit in salmonids. *Comparative Biochemistry and Physiology*, **107B**, 255–9.

Lau, K.R. (1985). The effects of salinity adaptation on intracellular chloride accumulation in the European flounder. *Biochimica et Biophysica Acta*, **818**, 105–8.

Laurent, P. (1984). Gill internal morphology. In *Fish Physiology. Vol 10A*, ed. W.S. Hoar & D.J. Randall, pp. 73–183. New York: Academic Press.

Laurent, P. & Dunel, S. (1980). Morphology of gill epithelia in fish. *American Journal of Physiology*, **238**, R147–59.

Laurent, P., Dunel-Erb, S., Chevalier, C. & Lignon, J. (1994). Gill epithelial cell kinetics in a freshwater teleost, *Oncorhynchus mykiss* during adaptation to ion-poor water and hormonal treatments. *Fish Physiology and Biochemistry*, **13**, 353–70.

Lingrel, J.B. & Kuntzweiler, T. (1994). Na^+,K^+-ATPase. *Journal of Biological Chemistry*, **269**, 19659–62.

Loretz, C.A., Freel, R.W. & Bern, H.A. (1983). Specificity of response of intestinal ion transport systems to a pair of natural peptide hormone analogues: somatostatin and urotensin II. *General and Comparative Endocrinology*, **52**, 198–206.

Loretz, C.A., Howard, M.E. & Siegel, A.J. (1985). Ion transport in goby intestine: cellular mechanism of urotensin II stimulation. *American Journal of Physiology*, **249**, G284–93.

Lucking, K., Jorgenson, P.L. & Meng, L.-M. (1994). Cloning and characterization of a new member of the family of Na^+/K^+-ATPase genes. In *The Sodium Pump*, ed. E. Bamberg & W. Schoner, pp. 53–56. Damstadt, Germany: Steinkopf Verlag.

Luke, G.A., Cutler, C.P., Sanders, I.L., Hazon, N. & Cramb, G. (1994). Branchial Na^+/K^+-ATPase expression in the European eel (*Anguilla anguilla*) following saltwater acclimation. In *The Sodium Pump*, ed. E. Bamberg & W. Schoner, pp. 246–249. Damstadt, Germany: Steinkopf Verlag.

MacKay, W.C. & Janicki, R. (1979). Changes in the eel intestine during seawater adaptation. *Comparative Biochemistry and Physiology*, **62A**, 757–61.

Madsen, S.S. (1990*a*). The role of cortisol and growth hormone in seawater adaptation and development of hypoosmoregulatory mechanisms in sea trout parr (*Salmo trutta*). *General and Comparative Endocrinology*, **79**, 1–11.

Madsen, S.S. (1990*b*). Cortisol treatment improves the development of hypoosmoregulatory mechanisms in the euryhaline rainbow trout, *Salmo gairdneri. Fish Physiology and Biochemistry*, **8**, 45–52.

Madsen, S.S. (1990*c*). Enhanced hypoosmoregulatory response to growth hormone after cortisol treatment in immature rainbow trout, *Salmo gairdneri. Fish Physiology and Biochemistry*, **8**, 271–9.

Madsen, S.S. & Naarmanse, E.T. (1989). Plasma ionic regulation and gill Na^+/K^+-ATPase changes during rapid transfer to seawater of yearling rainbow trout *Salmo gairdneri*: time course and seasonal variation. *Journal of Fish Biology*, **34**, 829–40.

Maetz, J. & Skadhauge, E. (1968). Drinking rates and gill ionic turnover in relation to external salinities in the eel. *Nature*, **217**, 371–3.

Mainoya, J.R. (1982). Water and NaCl absorption by the intestine of the tilapia *Sarotherodon mossambicus* adapted to fresh water or seawater and the possible role of prolactin and cortisol. *Journal of Comparative Physiology*, **146**, 1–7.

Mainoya, J.R. & Bern, H.A. (1984). Influence of vasoactive intestinal peptide and urotensin II on the absorption of water and NaCl by the anterior intestine of tilapia, *Sarotherodon mossambicus. Zoological Science*, **1**, 100–5.

Marshall, J., Martin, K.A., Picciotto, M., Hockfield, S., Nairn, A.C. & Kaczmarek, L.K. (1991). Identification and localisation of a dogfish homologue of human cystic fibrosis transmembrane conductance regulator. *Journal of Biological Chemistry*, **266**, 22749–54.

Marshall, W.S. (1981*a*). Sodium dependency of active chloride transport across isolated fish skin (*Gillichthys mirabilis*). *Journal of Physiology*, **319**, 165–78.

Marshall, W.S. (1981*b*). Active transport of Rb^+ across skin of the teleost *Gillichthys mirabilis. American Journal of Physiology*, **241**, F482–6.

Marshall, W.S. & Bern, H.A. (1979). Teleostean urophysis: urotensin II and ion transport across the isolated skin of a marine teleost. *Science*, **204**, 519–21.

Marshall, W.S. & Bern, H.A. (1980). Ion transport across the isolated skin of the teleost *Gillichthys mirabilis*. In *Epithelial Transport in the Lower Vertebrates*, ed. B. Lahlou, pp. 337–350. Cambridge: Cambridge University Press.

Marshall, W.S. & Bern, H.A. (1981). Active chloride transport by the skin of a marine teleost is stimulated by urotensin I and inhibited by urotensin II. *General and Comparative Endocrinology*, **43**, 484–91.

Marshall, W.S., Bryson, S.E. & Garg, D. (1993). α2-adrenergic-inhibition of Cl^- transport by opercular epithelium is mediated by intracellular Ca^{2+}. *Proceedings of the National Academy of Sciences, USA*, **90**, 5504–8.

Marshall, W.S., Bryson, S.E., Midelfart, A. & Hamilton, W.F. (1995). Low-conductance anion channel activated by cAMP in teleost Cl^--secreting cells. *American Journal of Physiology*, **268**, R963–9.

Marshall, W.S. & Nishioka, R.S. (1980). Relation of mitochondria-rich chloride cells to active chloride transport in the skin of a marine teleost. *Journal of Experimental Zoology*, **214**, 147–56.

May, S.A., Baratz, K.H., Key, S.A. & Degnan, K.J. (1984). Characterization of the adrenergic receptors regulating chloride secretion by the opercular epithelium. *Journal of Comparative Physiology*, **154**, 343–8.

May, S.A. & Degnan, K.J. (1984). cAMP-mediated regulation of chloride secretion by the opercular epithelium. *American Journal of Physiology*, **246**, R741–6.

May, S.A. & Degnan, K.J. (1985). Converging adrenergic and cholinergic mechanisms in the inhibition of C1 secretion in fish opercular epithelium. *Journal of Comparative Physiology*, **156**, 183–9.

Mayer-Gostan, N. & Lemaire, S. (1991). Measurements of gill ATPases using microplates. *Comparative Biochemistry and Physiology*, **98B**, 323–6.

McCormick, S.D. (1990*a*). Fluorescent labelling of Na^+,K^+-ATPase in intact cells by use of a fluorescent derivative of ouabain: salinity and teleost chloride cells. *Cell and Tissue Research*, **260**, 529–33.

McCormick, S.D. (1990*b*). Cortisol directly stimulates differentiation of chloride cells in tilapia opercular membrane. *American Journal of Physiology*, **259**, R857–63.

McCormick, S.D. & Bern, H.A. (1989). *In vitro* stimulation of Na^+–K^+-ATPase activity and ouabain binding by cortisol in coho salmon gill. *American Journal of Physiology*, **256**, R707–15.

Mendelson, S.A., Cherksey, B.D. & Degnan, K.J. (1981). Adrenergic regulation of chloride secretion across the opercular epithelium: the role of cAMP. *Journal of Comparative Physiology*, **145**, 29–35.

Mori, Y. & Ando, M. (1991). Regulation of ion and water transport across the eel intestine: effects of acetylcholine and serotonin. *Journal of Comparative Physiology B*, **161**, 387–92.

Morley, M., Chadwick, A. & El Tounsy, E.M. (1981). The effect of prolactin on water absorption by the intestine of the trout (*Salmo gairdneri*). *General and Comparative Endocrinology*, **44**, 64–8.

Musch, M.W., Orellana, S.A., Kimberg, L.S., Field, M., Halm, D.R., Krasny, E.J.Jr. & Frizzell, R.A. (1982). Na^+–K^+-Cl^- co-transport in the intestine of a marine teleost. *Nature*, **300**, 351–3.

Nagashima, K. & Ando, M. (1994). Characterisation of esophageal desalination in the seawater eel *Aguilla japonica. Journal of Comparative Physiology B*, **164**, 47–54.

Nonnotte, G. & Boeuf, G. (1995). Extracellular ionic and acid-base adjustments of atlantic salmon presmolts in fresh water and after transfer to sea water: the effects of ovine growth hormone on the acquisition of euryhalinity. *Journal of Fish Biology*, **46**, 563–77.

Nonnotte, L., Nonnotte, G. & Leray, C. (1986). Morphological changes in the middle intestine of the rainbow trout, *Salmo gairdneri*, introduced by a hyperosmotic environment. *Cell and Tissue Research*, **243**, 619–28.

Ogawa, M. (1974). The effects of bovine prolactin, sea water and environmental calcium on water influx in isolated gills of the euryhaline teleosts, *Anguilla japonica* and *Salmo gairdneri. Comparative Biochemistry and Physiology*, **49A**, 545–53.

Ogawa, M. (1975). The effects of prolactin, cortisol and calcium-free environment on water influx in isolated gills of Japanese eel, *Anguilla japonica. Comparative Biochemistry and Physiology*, **52A**, 539–43.

O'Grady, S.M., Field, M., Nash, N.T. & Rao, M.C. (1985). Atrial natriuretic factor inhibits Na-K-Cl cotransport in teleost intestine. *American Journal of Physiology*, **249**, C531–4.

Oide, M. & Utida, S. (1967). Changes in water and ion transport in isolated intestines of the eel during salt adaptation and migration. *Marine Biology*, **1**, 102–6.

Pagliarani, A., Ventrella, V., Ballestrazzi, R., Trombetti, F., Pirini, M. & Trigari, G. (1991). Salinity-dependence of the properties of gill (Na^++K^+)-ATPase in rainbow trout (*Oncorhynchus myskiss*). *Comparative Biochemistry and Physiology*, **100B**, 229–36.

Parmalee, J.T. & Renfro, J.L. (1983). Esophageal desalination of seawater in flounder: role of active sodium transport. *American Journal of Physiology*, **245**, R888–93.

Payan, P., Mayer-Gostan, N. & Pang, P.K.T. (1981). Site of calcium uptake in the fresh water trout gill. *Journal of Experimental Zoology*, **216**, 345–7.

Payne, J.A. & Forbush III, B. (1994). Alternatively spliced isoforms of the putative renal Na^+–K^+-Cl^- cotransporter are differentially distributed within the rabbit kidney. *Proceedings of the National Academy of Sciences, USA*, **91**, 4544–8.

Perrott, M.N., Grierson, C.E., Hazon, N. & Balment, R.J. (1992). Drinking behaviour in sea water and fresh water teleosts, the role of the renin–angiotensin system. *Fish Physiology and Biochemistry*, **10**, 161–8.

Pfeiler, E. & Kirschner, L.B. (1972). Studies on gill ATPase of rainbow trout (*Salmo gairdneri*). *Biochimica et Biophysica Acta*, **282**, 301–10.

Pickford, G.E., Pang, P.K.T., Weinstein, E., Torretti, J., Hendler, E. & Epstein, F.H. (1970). The response of the hypophysectomised cyprinodont, *Fundulus heteroclitus*, to replacement therapy and cortisol: effects on blood serum and sodium-potassium activated adenosine triphosphatase in the gills, kidney, and intestinal mucosa. *General and Comparative Endocrinology*, **14**, 524–34.

Pisam, M., Auperin, B., Prunet, P., Rentier-Delrue, F., Martial, J. & Rambourg, A. (1993). Effects of prolactin on α and β chloride cells in the gill epithelium of the seawater adapted tilapia *'Oreochromis niloticus'*. *Anatomical Record*, **235**, 275–84.

Pisam, M., Boeuf, G., Prunet, P. & Rambourg, A. (1990). Ultrastructural features of mitochondria-rich cells in stenohaline freshwater and seawater fishes. *American Journal of Anatomy*, **187**, 21–31.

Pisam, M., Caroff, A. & Rambourg, A. (1987). Two types of chloride cells in the gill epithelium of a freshwater-adapted euryhaline fish: *Lebistes reticulatus*; their modifications during adaptation to seawater. *American Journal of Anatomy*, **179**, 40–50.

Pisam, M., Chretien, M., Rambourg, A. & Clermont, Y. (1983). Two anatomical pathways for the renewal of surface glycoproteins in chloride cells of fish gills. *Anatomical Record*, **207**, 385–97.

Pisam, M., Prunet, P. & Rambourg, A. (1989). Accessory cells in the gill epithelium of the freshwater rainbow trout *Salmo gairdneri*. *American Journal of Anatomy*, **184**, 311–20.

Pisam, M. & Rambourg, A. (1991). Mitochondria-rich cells in the gill epithelium of teleost fishes: an ultrastructural approach. *International Review of Cytology*, **130**, 191–232.

Pisam, M., Sardet, C. & Maetz, J. (1980). Polysaccharidic material in chloride cell of teleostean gill: modifications according to salinity. *American Journal of Physiology*, **238**, R213–18.

Pressley, T.A. (1992). Phylogenetic conservation of isoform-specific regions within α-subunit of Na^+–K^+-ATPase. *American Journal of Physiology*, **262**, C743–51.

Prunet, P., Pisam, M., Claireaux, J.P., Boeuf, G. & Rambourg, A. (1994). Effects of growth hormone on gill chloride cells in juvenile Atlantic salmon (*Salmo salar*). *American Journal of Physiology*, **266**, R850–7.

Richman III, H.N. & Zaugg, W.S. (1987). Effects of cortisol and growth hormone on osmoregulation in pre- and desmoltified coho salmon (*Oncorhynchus kisutch*). *General and Comparative Endocrinology*, **65**, 189–98.

Riordan, J.R., Rommens, J.M., Kerem, B., Alon, N., Rozmahel, R., Grzelczak, Z., Zielenski, J., Lok, S., Plavsic, N., Chou, J.-L., Drumm, M.L., Iannuzzi, M.C., Collins, F.S. & Tsui, L.C. (1989). Identification of the Cystic Fibrosis gene: cloning and characterization of complementary DNA. *Science*, **245**, 1066–73.

Rudolph, H.K., Antebi, A., Fink, G.R., Buckley, C.M., Dorman, T.E., LeVitre, J.-A., Davidow, L.S., Mao, J.-I. & Moir, D.T. (1989). The yeast secretory pathway is perturbed by mutations in PMR1, a member of a Ca^{2+} ATPase family. *Cell*, **58**, 133–45.

Sakamoto, T., McCormick, S.D. & Hirano, T. (1993). Osmoregulatory actions of growth hormone and its mode of action in salmonids: a review. *Fish Physiology and Biochemistry*, **11**, 1–6.

Sardet, C., Pisam, M. & Maetz, J. (1979). The surface epithelium of teleostean fish gills. Cellular and junctional adaptations of the chloride cell in relation to salt adaptation. *Journal of Cell Biology*, **80**, 96–117.

Sargent, J.R. & Thomson, A.J. (1974). The nature and properties of the inducible sodium-plus-potassium ion-dependent adenosine triphosphatase in the gills of eels (*Anguilla anguilla*) adapted to fresh water and sea water. *Biochemical Journal*, **144**, 69–75.

Sargent, J.R., Thomson, A.J. & Bornancin, M. (1975). Activities and localization of succinic dehydrogenase and Na^+/K^+-activated adenosine triphosphatase in the gills of fresh water and sea water eels (*Anguilla anguilla*). *Comparative Biochemistry and Physiology*, **51B**, 75–9.

Scheer, B.T. & Langford, R.W. (1976). Endocrine effects on the cation-dependent ATPases of the gills of European eels (*Anguilla anguilla* L.) and efflux of Na. *General and Comparative Endocrinology*, **30**, 313–26.

Scheide, J.I. & Zadunaisky, J.A. (1988). Effect of atriopeptin II on isolated opercular epithelium of *Fundulus heteroclitus*. *American Journal of Physiology*, **254**, R27–32.

Schonrock, C., Morley, S.D., Okawara, Y., Lederis, K. & Richter, D. (1991). Sodium and potassium ATPase of the teleost fish *Catostomus commersoni*. Sequence, protein structure and evolutionary conservation of the α-subunit. *Biological Chemistry Hoppe-Seyler*'s, **372**, 279–86.

Seddiki, H., Maxime, V., Boeuf, G. & Peyraud, C. (1995). Effects of growth hormone on plasma ionic regulation, respiration and extracellular acid–base status in trout (*Oncorhynchus mykiss*) transferred to seawater. *Fish Physiology and Biochemistry*. (In press).

Shamraj, O.I. & Lingrel, J.B. (1994). A putative fourth Na^+,K^+-ATPase α-subunit gene is expressed in testis. *Proceedings of the National Academy of Sciences, USA*, **91**, 12952–6.

Shehadeh, Z.H. & Gordon, M.S. (1969). The role of salinity in adaptation of the rainbow trout, *Salmo gairdneri. Comparative Biochemistry and Physiology*, **30**, 397–418.

Silva, P., Solomon, R., Spokes, K. & Epstein, F.H. (1977). Ouabain inhibition of gill Na-K-ATPase: relationship to active chloride transport. *Journal of Experimental Zoology*, **199**, 419–26.

Simonneaux, V., Barra, J.A., Humbert, W. & Kirsch, R. (1987*a*). The role of mucus in ion absorption by the oesophagus of the sea-water eel (*Anguilla anguilla* L.). *Journal of Comparative Physiology*, **157**, 187–99.

Simonneaux, V., Humbert, W. & Kirsch, R. (1987*b*). Mucus and intestinal ion exchanges in the seawater adapted eel, *Anguilla anguilla* L. *Journal of Comparative Physiology*, **157**, 295–306.

Skadhauge, E. (1969). The mechanism of salt and water absorption in the intestine of the eel (*Anguilla anguilla*) adapted to waters of various salinities. *Journal of Physiology*, **204**, 135–58.

Skadhauge, E. (1974). Coupling of tansmural flows of NaC1 and water in the intestine of the eel (*Anguilla anguilla*). *Journal of Experimental Biology*, **60**, 535–46.

Suvitayavat, W., Dunham, P.B., Haas, M. & Rao, M.C. (1994). Characterisation of the proteins of the intestinal Na^+–K^+-$2Cl^-$ cotransporter. *American Journal of Physiology*, **267**, C375–84.

Tierney, M.L., Luke, G., Cramb, G. & Hazon, N. (1995). The role of the renin angiotensin system in the control of blood pressure and drinking in the European eel, *Anguilla anguilla. General and Comparative Endocrinology*. (In press).

Towle, D.W., Gilman, M.E. & Hempel, J.D. (1972). Rapid modulation of gill Na^++K^+-dependent ATPase activity during acclimation of the killifish *Fundulus heteroclitus* to salinity change. *Journal of Experimental Zoology*, **202**, 179–86.

Trischitta, F., Denaro, M.G., Faggio, C. & Schettino, T. (1989). Evidence for Na–K-Cl cotransport in fresh-water adapted eel intestine. *Pflügers Archives*, **415**, S12.

Trischitta, F., Denaro, M.G., Faggio, C. & Schettino, T. (1992). Comparison of Cl^--absorption in the seawater- and freshwater-adapted eel, *Anguilla anguilla*: evidence for the presence of an Na-K-C1 cotransport system on the luminal membrane of the enterocyte. *Journal of Experimental Zoology*, **263**, 245–53.

Utida, S., Hirano, T., Oide, H., Ando, M., Johnson, D.W. & Bern, H.A. (1972). Hormonal control of the intestine and urinary bladder in teleost osmoregulation. *General and Comparative Endocrinology* Suppl. 3, 317–27.

Van Praag, D., Farber, S.J., Minkin, E. & Primor, N. (1987). Production of eicosanoids by the killifish gills and opercular epithelia and their effect on active transport of ions. *General and Comparative Endocrinology*, **67**, 50–7.

Vigne, P., Lopez Farre, A. & Frelin, C. (1994). Na^{+}–K^{+}-Cl^{-} cotransporter of brain capillary endothelial cells. *Journal of Biological Chemistry*, **269**, 19925–30.

Wendelaar Bonga, S.E. & van der Meij, C.J.M. (1989). Degeneration and death, by apoptosis and necrosis, of the pavement and chloride cells in the gills of the teleost *Oreochromiz mossambicus*. *Cell and Tissue Research*, **255**, 235–43.

Xu, J.C., Lytle, C., Zhu, T.T., Payne, J.A. Benz, E. Jr & Forbush III, B. (1994). Molecular cloning and functional expression of the bumetanide-sensitive Na–K-C1 cotransporter. *Proceedings of the National Academy of Sciences, USA*, **91**, 2201–5.

Yamamoto, M. & Hirano, T. (1978). Morphological changes in the esophangeal epithelium of the eel, *Anguilla anguilla*, during adaptation to seawater. *Cell and Tissue Research*, **192**, 25–38.

Yoshikawa, J.S., McCormick, S.D., Young, G. & Bern, H.A. (1993). Effects of salinity on chloride cells and Na^{+}, K^{+}-ATPase activity in the teleost *Gillichthys mirabilis*. *Comparative Biochemistry and Physiology*, **105A**, 311–17.

Young, R.M. & Lingrel, J.B. (1987). Tissue distribution of mRNAs encoding the α isoforms and β subunit of Na^{+}, K^{+}-ATPase. *Biochemical and Biophysical Research Communications*, **145**, 52–8.

Zadunaisky, J.A. (1984). The chloride cell: the active transport of chloride and the paracellular pathways. In *Fish Physiology. Vol 10B*, ed. W.S. Hoar & D.J. Randall, pp. 129–176. New York: Academic Press.

Zadunaisky, J.A., Cardona, S., Au, L., Roberts, D.M., Fisher, E., Lowenstein, B., Cragoe, E.J.Jr. & Spring, K.R. (1995). Chloride transport activation by plasma osmolarity during adaptation to high salinity of *Fundulus heteroclitus*. *Journal of Membrane Biology*, **143**, 207–17.

G. GOLDSPINK

Temperature adaptation: selective expression of myosin heavy chain genes and muscle function in carp

By combining physiological and morphological methods with available and emerging molecular biology methods, we can now begin to understand adaptation to altered environmental conditions at the gene level. Fish are ectotherms and are therefore expected to be directly affected by temperature changes in water temperature. However, in our work we have found that some species are not completely at the mercy of the environment, as they can rebuild their muscle contractile apparatus for low temperature or for warm temperature swimming. They achieve this by expressing different sets of myofibrillar protein isogenes at different environmental temperatures.

Ectothermy is sometimes considered to be a more primitive state than endothermy. These states can more correctly be regarded as evolutionary options and it depends whether economy is more important than high locomotory speed for the survival of the species. Certainly, the resulting extra performance at the higher body temperature is expensive, as this has to be maintained all the time in all mammals and birds except during hibernation. In ectotherms in a few cases, extra performance is achieved by retaining heat generated by the locomotory muscles using a countercurrent heat exchanger or retina system, the examples, the eye heater organ and the myotomal muscle in tuna. In carp and similar fish species the evolutionary strategy has not been to raise the tissue temperature. Instead, by expressing a different set of genes, the enzyme system is changed so that it is optimized to the tissue temperature, which remains essentially the same as the ambient temperature.

Fish provide good examples to illustrate the point that different isozymes (protein isoforms) have evolved to operate at different temperatures, as different species occupy well-defined thermal niches ranging from −1.5 °C at the poles to 45 °C for some species that live in geothermal springs. Most species of fish are restricted to a narrow thermal range, e.g. Antarctic fish are restricted to −1.5 to +4 °C.

Warming above this latter temperature results in death. This type of adaptation has occurred during evolution and has involved the alteration of the enzymes and other proteins to acquire the appropriate activity and thermal stability for that temperature. Indeed, there seems to have been a trade-off between higher activity and thermal stability of the enzyme systems involved. A few species are subjected to rapidly changing thermal conditions, for example, fish that live in tidal pools where the temperature in the sun may reach the high 20s but may drop by 10 °C when the colder ocean water floods the pool with the incoming tide. In this case, thermal stability has been retained but the calcium sensitivity of the myofibrils of these fish is very temperature dependent so that the contractility of the locomotory muscles is altered at the excitation/contraction coupling level (Sidell, Johnston & Goldspink, 1984). Therefore, different myosin and other contractile protein genes have evolved which enable fish to swim effectively in warm and cold waters and even function well in situations of fluctuating temperatures. However, for us one of the most interesting types of adaptation is that shown by fish such as the carp which enables them to rebuild their contractile system for cold or warm temperature swimming, as this results in improved swimming performance and economy.

Muscle performance and economy of movement

For an animal species to exploit and survive in a given thermal environment, it has to escape predators and obtain food. Thus muscle performance is crucial to the survival of the individual and the species. Mechanical properties of muscle fibres are highly temperature dependent (Bennett, 1985) and therefore any adaptation would be expected to involve changing the muscle mechanism. As mentioned, another important factor is the thermal stability of the proteins. In order to study the activity and thermal stability of the contractile mechanism of muscles from Antarctic fish and fish from a range of environmental temperatures, we isolated myofibrils from the muscles of these fish and measured the specific ATPase and the rate at which the ATPase activity was lost by heat denaturation. The specific myofibrillar ATPase was much higher in the muscles of Antarctic fish than tropical or temperate water fish at low temperature. However, if the myofibrils were warmed up, the ATPase site was much more thermolabile than that of the tropical fish (Johnston, Frearson & Goldspink, 1973; Johnston *et al.*, 1975*b*) and with different thermodynamic properties (Johnston & Goldspink, 1975). Thus there seems to be a 'trade-off' in the Antarctic fish: higher ATPase activity and hence higher muscle power output at low temperatures, but decreased thermostability. Both parameters have important implications from the point of view of survival and geographical distribution of a given species.

The same sort of trade-off was seen during temperature acclimation in the carp. Myofibrillar ATPase measurements indicated that the rebuilding of their contractile system results in myofibrils with different temperature optima and with different thermostability (Johnston, Davison & Goldspink, 1975*a*; Penney & Goldspink, 1979, 1981). Carp and some other pond fish (Heap, Watt & Goldspink, 1985) are eurythermal whilst salmon and trout are also stenothermal species and have an optimum temperature of about 13 °C with a lethal upper limit of about 18 °C (Nahhas, 1981). However, even salmonid fish such as the salmon and trout do adjust to seasonal changes to temperature to some extent but unlike carp they cannot survive wide changes in temperature. In the case of carp, they are found swimming in the garden pond at about 0 °C in the winter and in ponds up to 30 °C in the tropics. The changes in the myofibril system are reflected at the isolated muscle fibres (Johnston & Brill, 1984) and whole muscle levels (Heap & Goldspink, 1985) through to the whole animal level (Rome, Loughna & Goldspink, 1985). In particular, the force/velocity characteristics of the muscle are changed with the cold-adapted fish having a more shallow curve when measurements are carried out at a low temperature (Beddow, 1994). This means that the crossbridge in these fish is able to produce more power at the low temperature. This is borne out by measurement of swimming performance of warm and cold-acclimated carp. These data showed that the red muscle of cold-adapted fish were able to generate more power so that these fish can maintain higher swimming speeds using their aerobic red muscles before they have to recruit their white fibres which will fatigue rapidly (Rome *et al.*, 1985).

When economy is one of the main issues, it is important that the muscle enzyme activity (myosin ATPase) is tuned to the velocity at which the muscle is required to shorten for maximum thermodynamic efficiency. Therefore, the white as well as the red muscle genes have to change during temperature adaptation. Indeed, differential gene expression in response to altered temperature is seen in both types of muscle in carp and similar species. This is reflected in the improved power output of the red muscle of the low-temperature-adapted fish (Rome *et al.*, 1985). The red muscle is used for cruising, and an adjustment in the enzymatic efficiency apparently leads to improved thermodynamic efficiency as illustrated by oxygen consumption measurements of cold-adapted carp (Smit, Van der Berg & Kijn-Den Hartog, 1974).

The alteration in the muscle properties particularly in the myofibrillar ATPase activity suggested that the myosin crossbridges were being changed. The ATPase and the actin binding sites belong to the myosin heavy chains, which are the main determinants of muscle contractility (Reiser *et al.*, 1985). The myosin light chains have been shown to be involved in the transmission of force rather than the determination of crossbridge cycling rates (Lowey, Waller & Tyrbus, 1993). Therefore, as part of the 'fine tuning' it seems that isoforms of other contractile proteins are also being produced during temperature acclimation. As far as the myosin heavy chains are concerned, the evidence all suggests that during adaptation different genes are expressed at different environmental temperatures. The rate of ATP utilization differs between different myosin crossbridge types. Therefore, the type of myosin *hc* genes expressed determine not only the maximum velocity of contraction but the economy of movement (i.e. rate of fatigue). We therefore focused our attention on the genes that encode the myosin crossbridge (Gerlach *et al.*, 1990). As the ATPase site and the actin binding site are located on the S1 fragment of the myosin heavy chain, we have isolated the 5′ end of the coding sequences of some of the fish genes. In some cases we have also isolated the 5′ flanking sequences, so that we can see how the warm and cold myosin *hc* genes differ and understand the gene switching mechanism during temperature adaptation.

Cloning and characterization of myosin *hc* genes from carp

From our earlier biochemical and physiological studies, it became apparent that the contractility of muscle in carp was changed significantly during acclimation to different temperature. The possibilities were that the contractile proteins such as the myosin *hc*s could be changed by post-translational processing of the proteins *in situ* or that the myofibrils are rebuilt by producing new protein isoforms. Also, the possibility existed that this may be achieved by differential splicing of the same myosin *hc* gene(s) or by the expression of a different set of myosin *hc* genes for warm temperature or cold water swimming. To answer the above questions concerning the differential expression of myosin during temperature adaptation, we prepared a genomic carp library. This was then screened for myosin heavy chain gene fragments by using cDNA obtained for the coding region of mammalian myosin *hc* genes. Only two mammalian myosin *hc* cDNA were available apart from some 3′ untranslated sequences (which would not be expected to have any homology with genes in another species). Using moderate stringency conditions, it was possible to isolate sequences of which restriction mapping showed 28 non-overlapping clones. This indicated that there were up to 28 different myosin *hc* genes in fish (Gerlach *et al.*, 1990), twice the number as in mammals (Goldspink *et al.*, 1992) but not as many as in birds (Robbins *et al.*, 1986). The next step was to characterize the different genes, and this was done by extracting RNA from red and white muscle from warm and cold acclimated carp. In this work, specific gene probes were used in Northern analysis to determine which myosin *hc* isogenes are expressed in red and white muscle in warm and cold acclimated fish (Gerlach *et al.*, 1990; Turay, Gerlach & Goldspink, 1991). At the time, protein separation methods were not particularly sensitive for large proteins, and different isoforms of myosin *hc* might have more or less the same change and mass. However, using peptide mapping Watabe's group in Japan (Hwang, Watabe & Hashimoto, 1990) demonstrated different myosin *hc* (S1) isoforms in warm and cold acclimated carp. This shows that the expression of the different myosin *hc* genes, which our group has documented, is seen at the protein level, which was predicted by the physiological and biochemical changes.

The main questions now are why is a different subset of myosin heavy chain genes expressed at low temperature to that expressed at warm temperatures and in different types of muscle? As well as characterizing the different fish myosin genes (Ennion *et al.*, 1995 and Chapter 7 of this Seminar Series), we have studied the 5′ flanking sequence of one of these to identify elements that may be involved in the temperature response.

Are some genes switched on by decreasing temperature or do some genes remain activated whilst others become inactive as the temperatures decreases?

In this Symposium we have learnt about two genes that are expressed at low environmental temperatures; the antifreeze gene (Chapter 1) and the desaturase gene (Chapter 2). As biological processes decrease with deceasing temperature, it is difficult to think of gene expression being up-regulated as the temperature is lowered. However, it is known that most genes in most cell types are inactive most of the time. This is usually because they are methylated but there are also *trans*-repressor proteins that compete with *trans*-activator proteins and switch off the gene in question. It is therefore feasible that the structural gene becomes active at low temperatures because the repressor gene becomes

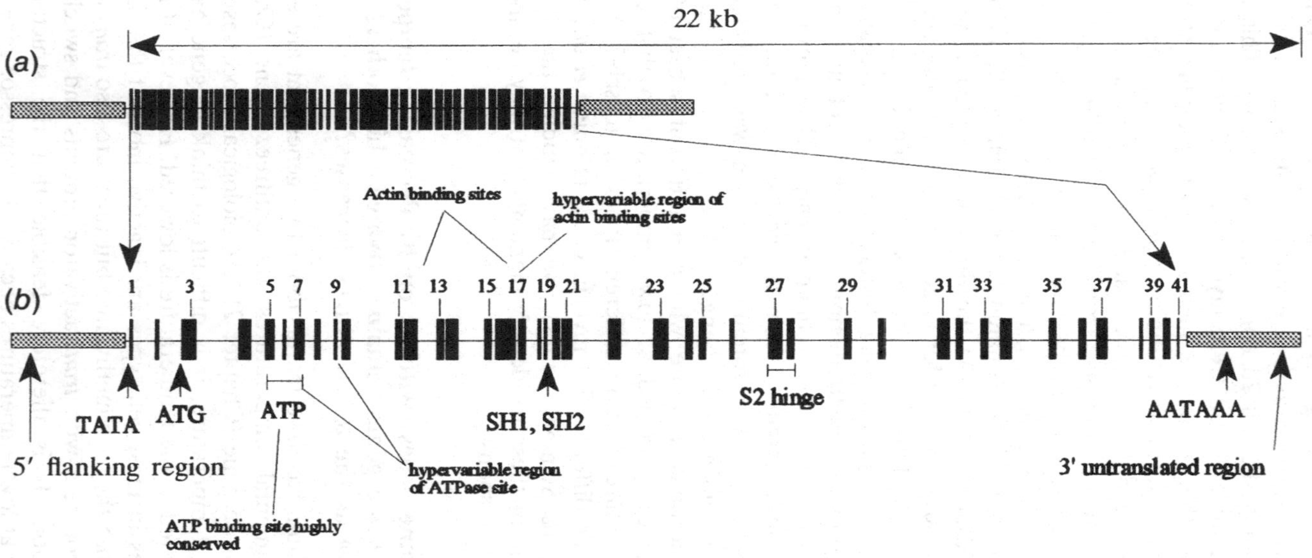

(*a*) Carp FG2 myosin heavy chain gene.

(*b*) 22 kb generalized mammalian myosin heavy chain gene.

inactive. In other words, the activation of these two genes and some of the myosin *hc* genes may be the result of positive induction of expression or more likely a negative effect involving de-repression. To answer these questions, we are studying the untranscribed and untranslated regions of myosin genes to identify the regulatory elements involved in myosin *hc* gene switching (Fig. 1).

Muscle gene families

In insects, only two myosin heavy chains have been described: a muscle form and a non-muscle form. The muscle myosin *hc* gene gives rise to a leg isoform and a wing isoform. Both of these are derived from the same gene by alternative splicing (Rozek & Davidson, 1983; Bernstein *et al.*, 1983; Kronert *et al.*, 1991). In the vertebrates the myosin *hc* genes have multiplied, resulting in families of genes, each one coding for a different myosin *hc* isoform. In mammals there are at least seven separate skeletal myosin *hc* genes. As well as being expressed in different tissues and in different cell types, different myosin *hc* genes including embryonic and neonatal are expressed at various stages during development (Butler-Brown & Whalen, 1984; Swynghedauw, 1986). *In toto* there are about 14 different myosin *hc* genes in the mammalian myosin *hc* gene superfamily. It is not surprising therefore that, in fish that are tetraploid, there are as many as 28 unique myosin genomic sequences (Gerlach *et al.*, 1990). This is a large enough gene family to permit some isoform genes to become specialized for expression in the different types of muscle: smooth, cardiac, and skeletal red, pink, white muscle at different developmental stages and for low temperature and warm temperature swimming.

In contrast to the myosin *hc* which are encoded by separate genes, the myosin light chains I and III are coded for by split genes or transcriptional units, but as yet these genes have only been studied in mammals and the

---

Fig. 1. A diagrammatic comparison of the fish FG2 myosin *hc* gene with a generalized mammalian myosin *hc* gene. The latter is about 22 kb whilst the fish gene is 12 kb. This is because the introns are shorter in the fish myosin *hc* gene, although the coding sequences are very homologous. In contrast, the flanking regions including the 5′ regulatory regions of the piscine gene bear very little resemblance to those of the mammalian gene except for the TATA, CCAAT and E boxes. As shown in Fig. 2, the E box is important in obtaining good expression, and constructs with this included appear to be temperature sensitive.

chicken. The different protein products of these split genes are obtained by splicing together different exon RNA transcripts so that several permutations are possible from the same gene. Thus, a single gene can give rise to several messenger RNAs and, hence, to different myosin light chain isoforms (Periasamy *et al.*, 1984). As far as the other myofibrillar proteins are concerned, actin is encoded by several genes, but only one alpha skeletal actin isoform is strongly expressed in adult skeletal muscle, although cardiac actin and cytoskeletal actin are expressed at low levels (Vandekerchove & Weber, 1979). The troponins and tropomyosins exist in several different isoforms in cardiac muscle and in fast and slow skeletal muscle. These are also derived by alternate splicing of the primary transcript of the troponin and of the tropomyosin genes during RNA processing (Breitbert *et al.*, 1985; Wieczorek, 1988).

The myosin *hc* genes are intrinsically interesting not only because muscle is the most abundant tissue in the body and myosin is the most abundant protein within muscle but because of their functional significance. The myosin *hc* genes occur as different isoforms (genes) which encode different types of myosin crossbridge (force generators). By comparing the cDNA sequences particularly that which encodes the ATPase and actin sites for the different myosin *hc* genes for different isoforms from different species, the strategic domains involved in the contractile process can be elucidated. Indeed, now that the three-dimensional structure is known for a chicken pectoralis muscle myosin S1 (Rayment *et al.*, 1993*a,b*) it should be possible to compare the three-dimensional configuration of the crossbridge head using computer graphics, by feeding in the data for other isoform genes of other species.

Determining cDNA sequences is now relatively straightforward and reasonably rapid using PCR methods. Therefore, a good number of smaller gene fragments in addition to the two cardiac myosin *hc* and some of the coding sequences of the skeletal and smooth muscle genes are available. A more difficult task is to clone and sequence the 5′ and 3′ flanking regions, which are more difficult to study as they are not transcribed or translated. Isolating the 5′ untranscribed/untranslated regions involved hybridizing a 5′ cDNA clone to a genomic library to identify clones with upstream sequences. These sequences are involved in controlling gene expression and therefore are of particular interest to anyone studying the regulation during adaptation and muscle phenotype determination.

Induction and repression of myosin heavy chain genes in mammals

Most genes studied have a promoter sequence upstream of the transcriptional start site, i.e. at the 5′ end. This is the sequence needed 'to drive' the gene. Experiments have shown that, without this promoter sequence, expression levels are very low. The binding of certain proteins or transcriptional factors at or near the promoter region results in activation or repression of the gene. The usual features of a promoter sequence are the presence of a TATA box that is believed to be the binding site for *cis*-acting factors and which is a few bases 'upstream' from the start site. This has been described as the 'signpost' for the RNA polymerase. Also, there is often a CAAT box which is thought to be the binding site for the *trans*-acting factor(s). Positive and negative elements are found to be present in mammalian myosin *hc* 5′ flanking regions, e.g. the response elements of thyroid hormone. The fast type isoform genes are activated by thyroid hormone, whilst the slow myosin *hc* genes are repressed. In the latter case, the repressor or negative site is just in front of the positive site and therefore the *trans*-activating proteins cannot bind when the negative site is occupied by a repressor protein. The *trans*-acting proteins aid or inhibit the attachment of the RNA polymerase and hence promote or repress gene transcription. Other regulatory sequences, which are called enhancers, are believed to facilitate the binding of the transcriptional factor proteins to DNA. These may be at the 5′ or 3′ end, and many bases from the start sequence. Although they may be remote in a linear manner, they are probably in close juxtaposition to the start sequence when the three-dimensional structure of the chromosome DNA is taken into account. Enhancers differ from promoters in that the orientation of the sequence does not seem to matter. Position is also not as crucial as for promoters. That is to say, the sequence can be reversed, yet the increase in the transcriptional level of the gene is still not altered. It is likely that enhancers facilitate the binding of *trans*-acting factors to the promoter of the gene, and thus increase the transcriptional rate of the gene in question.

Much research is now being directed to characterizing these transcriptional proteins which activate or repress gene expression. It appears that many genes have similar domains at the 5′ flanking region and therefore it is likely that the same *trans*-acting proteins activate or repress several related genes. Thyroid hormone (T3) activates transcription of skeletal fast myosin *hc* gene and the cardiac α-myosin *hc* gene. The receptor for this steroid hormone is the cytosolic proto-oncogene *c-erb* receptor, the protein product of which binds with the thyroid response element of the myosin *hc* gene sequence of the particular gene (Izumo, Nadal-Ginard & Mahdavi, 1986). There are also regulation elements in the 5′ flanking region of some of the myosin *hc* genes that represent binding sites for MyoD, myogenin and Myf-5, which determine the differentiation of mesenchymal cells into muscle cells during the embryonic differentiation of the tissue (Thayer *et al.*, 1989; Schafer *et al.*, 1990). A knowledge of the function of these regulatory sequences for key muscle genes and subsets of muscle genes is, therefore, essential if growth is to be regulated in a scientific manner by the manipulation of gene expression including the introduction of engineered genes.

Factors involved in the regulation of myosin hc genes

Again, we have to try to learn from what we know about mammalian muscle, which is a tissue in which gene expression is regulated to a large extent by physical signals. In mammalian muscle, changes in muscle phenotype which involve switching of myosin isoform gene expression are induced by acute stretch and electrical stimulation (Goldspink *et al.*, 1992). The inherent ability of mammalian skeletal muscle to adapt to mechanical signals is related to its ability to switch off different isoform genes and to alter the general levels of expression of different subsets of genes. Gene switching in mammalian muscle has also been shown to be influenced by hormones (Lompre, Nadal-Ginard & Mahdavi, 1984; Izumo *et al.*, 1986). In fish it was therefore feasible that temperature does not have a direct effect on gene expression switching and that changes in water viscosity or hormonal balance may induce myosin *hc* gene switching.

The details of the molecular mechanism(s) involved in isoform gene switching, particularly the links between the physical signals and the gene activation and repression are not known. Two possible mechanisms spring to mind including transient changes in internal calcium levels and metabolic signals such as the depletion of ATP. It may also involve the release of growth factors from muscle fibres, which then act in an autocrine fashion and cause the up-regulation of certain genes, particularly those associated with growth.

Regulation of carp myosin heavy chain genes

We have isolated and characterized a member of the carp myosin heavy chain gene family including its 5′ regulatory sequence. This myosin *hc* isogene is of particular interest as Northern and *in situ* hybridization studies showed that this gene is induced by increasing the environmental temperature, and is only expressed in the small white myotomal muscle fibres. In another fish species these small fibres have been shown to be associated with hyperplastic growth (Rowlerson *et al.*, 1995). Therefore, this particular myosin can be regarded as a growth gene (Gauvry *et al.*, 1994). The whole isoform gene, including potential regulatory sequence 5′ to the transcription start site and the 3′ untranslated region was cloned in a λ2001 bacteriophage vector. This was possible as the overall length of this and other fish genes is only about one-half of those in mammals and birds. This is due to shorter introns in the fish genes.

It was particularly interesting to compare the sequence of the fish gene with the available mammalian and avian myosin *hc* gene sequences. As might have been expected, 5′ coding region of the gene revealed high amino acid sequence homology with translated exons 3 to 7 of mammalian myosin heavy chain genes indicating identical exon/intron boundaries. In other words, the exons encoding the crossbridge had a high homology with those in similar myosin *hc* isogenes in mammals or birds. The situation with respect to the 5′ flanking region was very different. Although putative TATA and CCAAT boxes were found, the other parts showed sequence homology to the 5′ flanking regions of its mammalian and avian counterparts. This indicates that the mechanism of regulating gene expression may be somewhat different in ectothermal aquatic organisms to that in endotherms (Fig. 2).

To study the regulation of expression, different lengths of the 5′ flanking sequence were spliced to a reporter gene chloramphenicol acetyltransferase (CAT) sequence. The resulting genes, which consisted of the different lengths of the 5′ flanking sequence, were used to transfect muscle cells in culture or for direct injection into carp skeletal muscle (Hanson *et al.*, 1991). Skeletal and cardiac muscle cells have been shown to take up plasmid DNA when introduced by a simple intramuscular injection. However, the number of fibres transfected and the levels of expression increased when mammalian type viral promoters were used (Wolff *et al.*, 1990). Our group showed that, when the engineered gene was under the control of a muscle-specific promoter (Hansen *et al.*, 1991), and when the genes were injected into young muscle (Wells & Goldspink, 1992), expression increased levels were much higher. This therefore presented a means of assaying promoter activity of the 5′ fragments involved. The 5′ flanking region which contains a consensus sequence known as an 'E-box' (CANNTG) was shown to be important for the expression of the reporter gene. When fish acclimated at 18 °C or 20 °C were injected with the gene constructs containing this region (−901, +15), the expression level was about 500% greater than at 18 °C. Therefore, this appears to be the temperature-sensitive region.

Deletion constructs of the promoter region of the FG2 carp myosin heavy chain gene

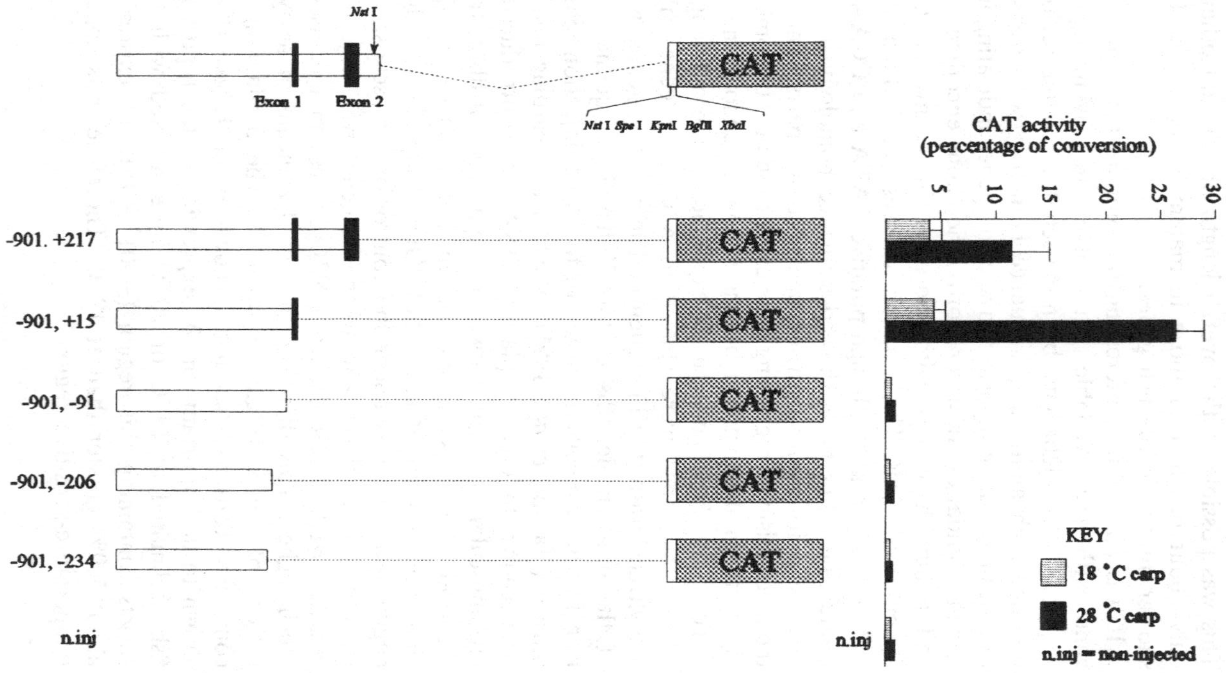

The way forward

By transferring engineered genes into fish and other species, many possibilities exist. However, in order to do this effectively, it is necessary to isolate and characterize appropriate promoter (regulatory) sequence so that good tissue specific expression is obtained. The promoter described above has already been used to generate transgenic carp and zebrafish (see Chapter 10). In theory, any coding sequence will be expressed, providing it is under the control of an appropriate promoter which, in this case, is a fish-specific muscle promoter. In this way, genes for growth enhancement and disease resistance can be introduced into the germ line. Fish can also be used as factories for producing biologically active proteins for pharmaceutical purposes. As a safer alternative than manipulating the germ line, similar genes can be introduced into individual fish in an inexpensive manner by the direct gene injection method described above. The injection procedure can be automated so vaccine DNA can be introduced very cost effectively. However, these applied aspects of fish molecular genetics have first to be underpinned by basic research into the mechanisms involved in the expression of individual genes, and the way they are, so that these developments can proceed in a logical, scientific fashion.

Fig. 2. This shows the sort of analyses that we have carried out (Gauvry *et al.*, 1995) in order to elucidate how the fish myosin *hc* genes are controlled and affected by environmental temperature. The approach was to splice regions of the 5′ flanking sequence of the FG2 gene to a chloramphenicol (CAT) reporter sequence to determine the levels of activity when transfected into a muscle cell line and when injected directly into carp myotomal muscle (Hansen *et al.*, 1991). Using the latter method, the reporter genes which included the promoter sequence −901 plus 15 base of the 5′ untranslated region (5′ UTR) gave the best expression. At 28 °C this was five times that at 18 °C, indicating that this is the temperature-sensitive response region. Further work is required to elucidate the mechanism by which environmental temperature markedly alters gene expression using the fast gene system.

References

Beddow, T.A. (1994). The influence and temperature acclimation on isolated muscle properties and burst swimming performance of the sculpin (*Myoxocephalus scorpius* L.). PhD thesis, The University of St Andrews.

Bennett, A.F. (1985). Temperature and muscle in design and performance of muscle systems. *Journal of Experimental Biology*, **115**, 333–44.

Bernstein, S.L., Mogami, K., Donady, J.J. & Emerson, C.P. Jr. (1983). Drosophilia muscle myosin heavy chain encoded by a single gene in a cluster of muscle mutations. *Nature*, **302**, 393–7.

Breitbert, R.E., Nguyen, H.T., Medford, R.M., Destree, A.T., Mahdavi, V. & Nadal-Ginard, B. (1985). Intricate combinational patterns of exon splicing generate multiple regulated troponin-T isoforms from a single gene. *Cell*, **41**, 67.

Butler-Brown, G.S. & Whalen, R.G. (1984). Myosin isozyme transitions occurring during the post-natal development of the rat soleus muscle. *Developmental Biology*, **102**, 324–34.

Crockford, T. & Johnston, I.A. (1990). Temperature acclimation and the expression of contractile protein isoforms in the skeletal muscle of the common carp (*Cyprinus carpio* L.). *Journal of Comparative Physiology (B)*, **160**, 23–30.

Ennion, S., Gauvry, L., Butterworth, P. & Goldspink, G. (1995). Small diameter white myotomal muscle fibres associated with growth hyperplasia in the carp (*Cyrinis carpio*) express a distinct myosin heavy chain gene. *Journal of Experimental Biology*, **198**, 1603–11.

Gauvry, L., Ennion, S., Hansen, E., Butterworth, P. & Goldspink, G. (1995). Cloning and characterisation of a carp myosin heavy chain gene promoter region. *European Journal of Biochemistry*, (In press).

Gauvry, L., Hansen, E., Ennion, S.J. & Goldspink, G. (1994). Differences in the transcribed and flanking regions of carp and mammalian myosin heavy chain genes. In preparation.

Gerlach, G., Turay, L., Malik, T.A., Lida, J., Scutt, A. & Goldspink, G. (1990). Mechanism of temperature acclimation in the carp: a molecular approach. *American Journal of Physiology*, **259**, R237–44.

Goldspink, G., Scutt, A., Loughna, P., Wells, D., Jaenicke, T. & Gerlach, G.-F. (1992). Gene expression in skeletal muscle in response to mechanical signals. *American Journal of Physiology*, **262**, R326–63.

Hansen, E., Goldspink, G., Butterworth, P.W. & Chang, K.C. (1991). Strong expression of some mammalian gene constructs in fish muscle during direct gene transfer. *FEBS Letters*, **290**, 73–6.

Heap, S.P., Watt, P.W. & Goldspink, G. (1985). Consequences of thermal change on the myofibrillar ATPase of five freshwater teleosts. *Journal of Fish Biology*, **26**, 733–8.

Heap, S.P., Watt, P.W. & Goldspink, G. (1987). Contractile properties of goldfish fin muscles following temperature acclimation. *Journal of Comparative Physiology, B*, **157**, 219–25.

Heap, S.P. & Goldspink, G. (1985). Alterations to the swimming performance of carp, *Cyprinus carpio*, as a result of temperature acclimation. *Journal of Fish Biology*, **29**, 747–53.

Hwang, G.C., Watabe, S. & Hashimoto, K. (1990). Changes in carp myosin ATPase induced by temperature acclimation. *Journal of Comparative Physiology, B*, **160**, 233–9.

Izumo, S., Nadal-Ginard, B. & Mahdavi, V.J. (1986). All members of the MHC multigene family respond to thyroid hormone in a highly tissue-specific manner. *Science*, **231**, 597–600.

Johnston, I.A., Davison, W. & Goldspink, G. (1975*a*). Adaptations in magnesium-activated myofibrillar ATPase activity induced by environmental temperature. *FEBS Letters*, **50**, 293–5.

Johnston, I.A., Frearson, N. & Goldspink, G. (1973). The effect of environmental temperature on the properties of myofibrillar ATPase from various species of fish. *Biochemical Journal*, **133**, 735–8.

Johnston, I.A., Davison, W., Walesby, N.J. & Goldspink, G. (1975*b*). Temperature adaptation in myosin of Antarctic fish. *Nature*, **254**, 74–5.

Johnston, I.A. & Brill, R. (1984). Thermal dependence of contractile properties of single skinned muscle fibres isolated from Antarctic and various Pacific marine fish. *Journal of Comparative Physiology, B*, **155**, 63–70.

Johnston, I.A. & Goldspink, G. (1975). Thermodynamic activation parameters of fish myofibrillar ATPase enzyme and evolutionary adaptations to temperature. *Nature*, **257**, 620–2.

Kronert, W.A., Edwards, K.A., Roche, E.S., Wells, L. & Bernstein, S.I. (1991). Muscle-specific accumulation of *Drosophila* myosin heavy chains: a splicing mutation in an alternative exon results in an isoform substitution. *EMBO Journal*, **10**, 2479–88.

Lompre, A.M., Nadal-Ginard, B. & Mahdavi, V.J. (1984). Expression of the cardiac ventricular α- and β-myosin heavy chain genes is developmentally and hormonally regulated. *Journal of Biological Chemistry*, **259**, 6437–46.

Lowey, S., Waller, G.S. & Tyrbus, K.M. (1993). Skeletal muscle myosin light chains are essential for physiological speeds of shortening. *Nature*, **365**, 454–6.

Nahhas, R. (1981). Studies of growth and swimming capability of young trout under controlled conditions. PhD, University of Hull.

Penney, R.K. & Goldspink, G. (1979). Compensation limits of fish muscle myofibrillar ATPase enzyme to environmental temperature. *Journal of Thermal Biology*, **4**, 269–72.

Penney, R.K. & Goldspink, G. (1981). Temperature adaptation by the myotomal muscle of fish. *Journal of Thermal Biology*, **6**, 297–306.

Periasamy, M., Strehler, E.E., Garfinkel, L.I., Gubits, R.M., Ruiz-Opazo, N. & Nadal-Ginard, B. (1984). Fast skeletal muscle myosin light chains 1 and 2 are produced from a single gene by a combined process of differential RNA transcription and splicing. *Journal of Biological Chemistry*, **259**, 13595.

Rayment, I., Holden, H.M., Whitaker, M., Yohn, C.B., Lorenz, M., Holmes, K.C. & Milligan, R.A. (1993*a*). Structure of the actin-myosin complex and its implication for muscle action. *Science*, **261**, 58–65.

Rayment, I., Rypniewski, W.R., Schmidt-Base, K., Smith, R., Tomchick, D.R., Benning, M.M., Winkelman, D.A., Wesenberg, G. & Holden, H.M. (1993*b*). Three-dimensional structure of myosin subfragment 1: A. *Molecular Motor Science*, **261**, 50–8.

Reiser, P.J., Moss, R.L., Giulian, G.C. & Greaser, M.L. (1985). Shortening velocity and myosin heavy chains of developing rabbit muscle fibres. *Journal of Biological Chemistry*, **206**, 14403–5.

Robbins, J., Horan, R., Gulick, J. & Kropp, K. (1986). The chicken myosin heavy chain family. *Journal of Biological Chemistry*, **261**, 6606–12.

Rome, L.C., Loughna, P.T. & Goldspink, G. (1985). Temperature acclimation: improved sustained swimming performance in carp at low temperatures. *Science*, **228**, 194–6.

Rowlerson, A., Mascarello, F., Radaelli, G. & Veggti, A. (1995). Differentiation and growth of muscle in the fish *Sparus aurata* (L.). II. Hyperplastic and hypertrophic growth of lateral muscle from hatching to adult. *Journal of Muscle Research and Cell Motility*, **16**, 213–22.

Rozek, C.E. & Davidson, N. (1983). Drosophila has one myosin heavy-chain gene with three developmentally regulated transcripts. *Cell*, **32**, 23–34.

Schafer, B.W., Blakely, B.T., Darlington, G.J. & Blau, H.M. (1990). Effect of cell history on response to helix–loop–helix family of myogenic regulators. *Nature*, **344**, 454–8.

Schmidt, G.R., Goldspink, G., Roberts, R., Katenschmidt, L.L., Cussens, R.G. & Briskey, E.J. (1972). Electromyography and resting membrane potential in longissimus muscle of stress–susceptible and stress-resistant pigs. *Journal of Amino Acid Science*, **34**, 379–83.

Sidell, B.D., Johnston, I.A. & Goldspink, G. (1984). The eurythermal myofibrillar protein complex of the numichog. (*Fundulus heteroclitus*): adaptation to a fluctuating thermal environment. *Journal of Comparative Physiology*, **153**, 167–73.

Smit, H., Van der Berg, R.J. & Kijn-Den Hartog, I. (1974). Some experiments on thermal acclimation in the goldfish *Carassius auratus*. *Netherlands Journal of Zoology*, **24**, 32–49.

Swynghedauw, D. (1986). Developmental and functional adaptations of contractile proteins in cardiac and skeletal muscles. *Physiological Reviews*, **65**, 710–71.

Thayer, M.J., Tapscott, S.J., Davis, R.L., Wright, W.E., Lassar, A.B. & Weintraub, H. (1989). Positive autoregulation of the myogenic determination gene MyoD1. *Cell*, **58**, 241–8.

Turay, L., Gerlach, D.-F. & Goldspink, G. (1991). Changes in myosin *hc* gene expression in the common carp during acclimation to warm environmental temperatures. *Journal of Physiology*, **435**, 102p.

Vandekerchove, J. & Weber, K. (1979). The complete amino acid sequence of actins from bovine heart, bovine skeletal muscle and rabbit slow skeletal muscle. *Differentiation*, **14**, 123.

Wells, D.J. & Goldspink, G. (1992). Age and sex influence expression of plasmid DNA directly injected into mouse skeletal muscle. *FEBS Letters*, **306**, 203–5.

Wieczorek, D.F. (1988). Regulation of alternatively sliced and tropomyosin gene expression by nerve extract. *Journal of Biological Chemistry*, **263**, 10456–9.

Wolff, J.A., Malone, R.W., Williams, P., Chong, W., Ascasadi, G., Jani, A. & Felgner, P.L. (1990). Direct gene transfer into mouse muscle *in vivo*. *Science*, **247**, 1465–8.

A.J. EL HAJ

Crustacean genes involved in growth

Introduction

Crustaceans represent a group of animals which grow intermittently throughout juvenile and adult life. Development involves a series of larval stages before metamorphosing into the young juvenile. In the adult, growth is centred around ecdysis when the old exoskeleton is lost and the new exoskeleton expands by a subsequent increase in size. The number of moults can depend on many factors, both environmental and intraspecific with up to 12 moults in the first year of life. Increments in crabs can vary by 3–44% of the original carapace width (Hartnoll, 1982) with growth completed over 2–7 days. In some cases growth can be biphasic, for example, in the isopod, *Idotea rescata*, with the posterior half of the animal shedding prior to the anterior half. In the edible crab, *Cancer pagurus*, it takes up to 15–20 years for an adult male to reach maximal size (Bennet, 1974) whereas the common green shore crab, *Carcinus maenas*, takes only 4 years (Crothers, 1967).

Tissue growth has been shown to occur in cycle with increasing body size; the predominant growth phase occurring during the pre- and post-moult phases. This is also true for the major muscle groups such as the abdominal and leg muscles (El Haj, Govind & Houlihan, 1984). Crustacean muscle is similar in structure to vertebrate and mammalian skeletal muscle with muscle proteins assembled into sarcomeric units aligned along a fibre. Muscles are composed of large fibres arranged in a pennate fashion around a central apodeme. A major difference between vertebrate and crustacean muscle fibres is the length of the sarcomeres which varies with fibre type; fast muscle composed of short sarcomeres with low mitochondrial density, slow fibres with longer sarcomeres and high mitochondrial density and intermediate fibres in-between (Fig. 1). There is evidence that further differentiation of fibre types exists within these three broad groupings and a number of different isoforms exist for myosin heavy chain (Li & Mykles, 1990).

Hormonal regulation of the moult and tissue growth has been the subject of numerous studies, and the structure and function of the

(a)

(b)

(c)

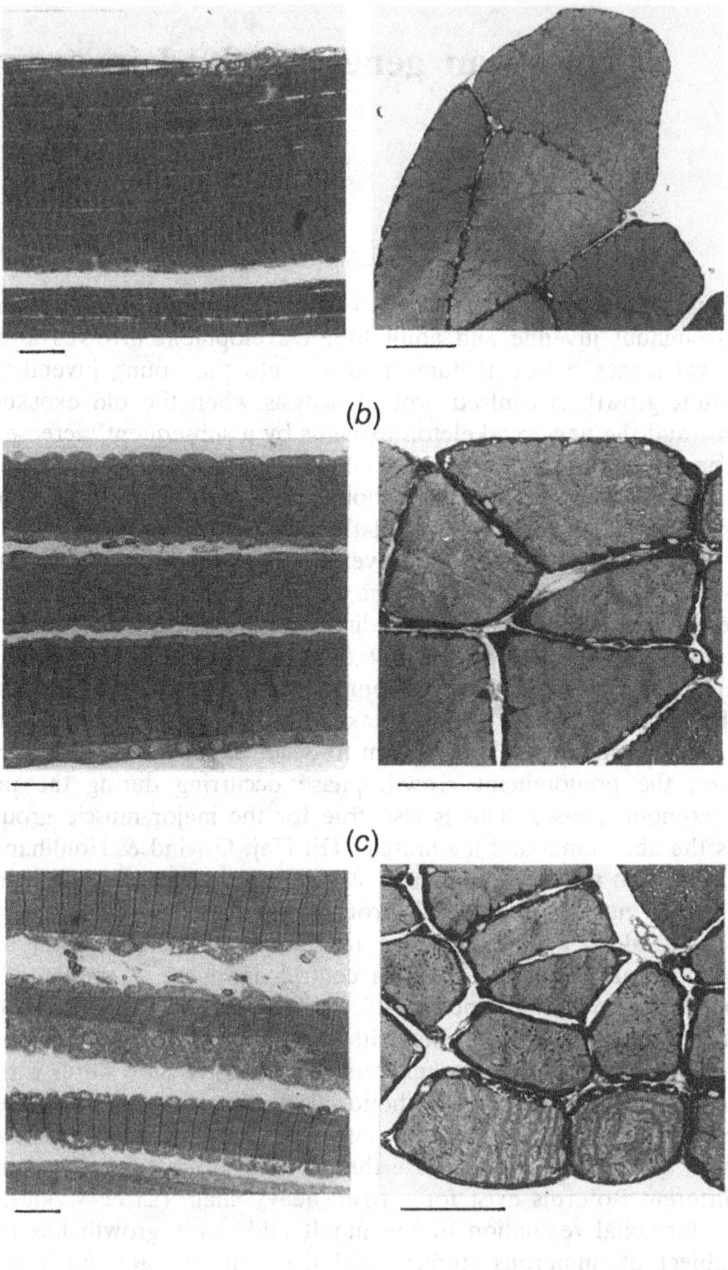

moult/growth regulatory hormones are becoming increasingly well characterized (Chang, 1993). A finely tuned interaction between the moult-inhibiting hormone (MIH) produced by the x-organ–sinus gland complex and the moult-promoting hormones produced by the y-organs, e.g. the ecdysteroids and ponasterone A, act to regulate the moult cycle. In addition, crustacean hyperglycaemic hormone (CHH) produced by the x-organ–sinus gland complex may be involved in the interaction between the moult and circadian rhythms, and methyl farnosoate has been implicated in the control of reproductive processes. The expression of receptors in specific tissues, and our understanding of receptor activity in relation to the moult cycle, have also been studied (El Haj, Harrison & Chang, 1994; Webster & Keller, 1988).

In mammals, the majority of myofibrillar protein isoforms have been cloned, e.g. the heavy and light chain myosins, troponins, tropomyosins and actin (Buckingham & Minty, 1983; Bandman, 1992). In the arthropods, the genes involved in muscle development in *Drosophila* have been cloned, and much work has been carried out identifying mutations and the effects of mutations on muscle assembly or function (Bernstein, O'Donnell & Cripps, 1993; Sparrow *et al.*, 1992). Over the past five years, the major sarcomeric protein and some of the regulatory hormone genes in Crustacea have been cloned and used to study the mRNA distribution in the tissues. These genes have been cloned in species of Malacostracan crusatceans including Isopoda, Brachyura and Natantia (Weidermann, Gromoll & Keller, 1989; Ortega *et al.*, 1992; Harrison & El Haj, 1994; Cotton & Mykles, 1994; De Kleijn *et al.*, 1994).

Actin

In the past it has been possible to take advantage of the strong homologies in some proteins to enable using heterologous probes from

---

Fig. 1. Photomicrograph of crustacean extensor muscle from the walking leg. Sections show the muscle fibres in longitudinal and cross-section. Note the differences in sarcomere length and mitochondrial density: (*a*) fast phasic fibres with short sarcomeres and low mitochondrial density, (*b*) intermediate fibres (*c*) slow tonic fibres with long sarcomeres and high mitochondrial density. Transverse sections are stained with *p*-phenylenediamine which stains the mitochondria located in the periphery of the fibres.
Scale L.S. = 25 μm; T.S. = 1 mm.

other species. In particular, a mammalian cDNA for actin has been used to study the regulation of actin mRNA levels in *Carcinus maenas* and *Austropotamobius pallipes*. El Haj, Harrison & Whiteley (1992) used a mammalian actin cDNA (Minty *et al.*, 1981) for hybridization with total RNA extracted from muscle from different species of Crustacea identifying two mRNAs for actin, 1.65 and 2.1 kb. Using the sequence data from this clone and a corresponding *Drosophila* clone, a pair of degenerate primers were devised and used to amplify a 736 bp species of cDNA clone from *Homarus gammarus*. The primers were chosen to include the majority of the coding region of the actin cDNA, excluding approximately 60 bases at the 5′ end and 2300 bases at the 3′ end (Harrison & El Haj, 1994). A restriction map of the clone and the nucleotide sequence/predicted amino acid sequence are shown in Figs. 2 and 3(*a*). Northern blots of total RNA from *Homarus gammarus* muscle and gonad tissue recognize an mRNA of 1.6 kb in size under high stringency washes (Fig. 3(*b*)).

Four cDNA clones coding for *Artemia* actin isoforms have been identified, which encode for sarcomeric and cytoplasmic actins in *Artemia* (Macias & Sastre, 1990). The number of actin genes in *Artemia* has been suggested to be eight to ten by Southern analysis which is larger than the six genes proposed for *Drosophila* (Fyberg *et al.*, 1980). In *Drosophila*, two of the genes encode cytoplasmic actins whereas four encode muscle-specific actin isoforms (Fyrberg *et al.*, 1983). Despite some heterogeneity of the structure of actin genes, the coding sequences of all the genes are highly conserved (Fyrberg *et al.*, 1981, 1983; Manseau, Ganetsky & Craig, 1988) (Table 1). The *N*-terminal amino acid sequence of the *Homarus* sp. clones shows similarities with other arthropod actins and cytoplasmic vertebrate actins (Table 1). Macias & Sastre (1990) also suggest that post-translational processing of actin may be similar in *Artemia* to other organisms as defined by the amino acid structure.

Myosin heavy chain

Sarcomeric myosin heavy chain consists of protein isoform multigene families in a range of organisms from nematodes to mammals. In the

---

Fig. 2. Nucleotide sequence and predicted amino acid sequence of the *H. gammarus* actin cDNA cloned in pHgAct. Lines above the sequences indicate the positions of the primers. Comparisons are made with ArAct302 (an *Artemia* actin clone) at the nucleotide and amino acid level (from Harrison & El Haj, 1994).

```
HgAct(nt)       1    T GGG TTC GCG GGG GAT GAT GCG CCT CGT GCT GTC TTC CCA TCC  43
HgAct(aa)                                              Arg Ala Val Phe Pro Ser

ArAct302(nt)    1  ... ... ... ... ... ... ... ... ... ... ... ... ... ... GC    2
                                                                           ::
HgAct(nt)      44  ATC GTG GGC CGA CCC CGC CAT CAG CGC GTG ATG GTG GGC ATG GGC  88
HgAct(aa)          Ile Val Gly Arg Pro Arg His Gln Gly Val Met Val Gly Met Gly

ArAct302(nt)    3  CAA AAA GAT AGC TAT GTC GGT GAT GAG GCT CAG AGC AAA CGT GGT  47
                   ::  ::  ::      ::  ::  ::: ::  ::: ::  ::: ::: ::   :  ::
HgAct(nt)      89  CAG AAG GAC TCG TAC GTA GGT GAC GAG GCA CAG AGC AAG AGA GGC  133
HgAct(aa)          Gln Lys Asp Ser Tyr Val Gly Asp Glu Ala Gln Ser Lys Arg Gly
ArAct302(aa)        -   -   -   -   -   -   -   -   -   -   -   -   -   -   -

ArAct302(nt)   48  ATT CTT ACC CTC AAA TAC CCA ATC GAG CAC GGT GTT GTC ACT AAC  92
                   ::  ::  ::: ::  ::: ::  ::  ::: ::: ::: :::  :: ::: ::  :::
HgAct(aa)     134  ATC CTC ACC CTC AAA TAT CCC ATC GAG CAC GGT ATT GTC ACC AAC  178
HgAct(aa)          Ile Leu Thr Leu Lys Tyr Pro Ile Glu His Gly Ile Val Thr Asn
ArAct302(nt)        -   -   -   -   -   -   -   -   -   -   -  Val  -   -   -

ArAct302(nt)   93  TGG GAT GAT ATG GAA AAG ATT TGG CAT CAT ACC TTT TAC AAT GAG  137
                   ::: ::  ::: ::: ::: ::: ::  ::: ::: ::  ::  ::  ::: ::: ::
HgAct(nt)     179  TGG GAC GAT ATG GAA AAG ATC TGG CAT CAC ACT TTC TAC AAT GAA  223
HgAct(aa)          Trp Asp Asp Met Glu Lys Ile Trp His His Thr Phe Tyr Asn Glu
ArAct302(aa)        -   -   -   -   -   -   -   -   -   -   -   -   -   -   -

ArAct302(aa)  138  CTT CGT GTT GCT CCA GAA GAA CAC CCC GTC CTC CTG ACA GAG GCT  182
                   ::  ::: ::: ::  ::: ::  ::  ::: ::: ::: ::   :: ::: ::: :::
HgAct(nt)     224  CTG CGT GTT GCC CCA GAG GAG CAC CCC GTC CTG TTG ACA GAG GCT  268
HgAct(aa)          Leu Arg Val Ala Pro Glu Glu His Pro Val Leu Leu Thr Glu Ala
ArAct302(aa)        -   -   -   -   -   -   -   -   -   -   -   -   -   -   -

ArAct302(nt)  183  CCC TTG AAC CCA AAA GCC AAT AGA GAA AAA ATG ACA CAA ATT ATG  227
                   ::: ::  ::: ::  ::  ::: ::   :  ::: ::  ::: ::  ::: ::: :::
HgAct(nt)     269  CCC CTC AAC CCT AAG GCC AAC CGT GAA AAG ATG ACC CAA ATT ATG  313
HgAct(aa)          Pro Leu Asn Pro Lys Ala Asn Arg Glu Lys Met Thr Gln Ile Met
ArAct302(aa)        -   -   -   -   -   -   -   -   -   -   -   -   -   -   -

ArAct302(nt)  228  TTT GAA ACC TTC AAC ACC CCA GCA ATG TAC GTT GCC ATT CAA GCT  272
                   ::  ::: ::  ::: ::: ::  ::  ::  ::: ::: ::  ::  ::  ::  :::
HgAct(nt)     314  TTC GAA ACA TTC AAC ACT CCC GCC ATG TAC GTC GCT ATC CAG GCT  358
HgAct(aa)          Phe Glu Thr Phe Asn Thr Pro Ala Met Tyr Val Ala Ile Gln Ala
ArAct302(aa)        -   -   -   -   -   -   -   -   -   -   -   -   -   -   -

ArAct302(nt)  273  GTT CTC TCG CTT TAT GCG TCA GGT CGT ACA ACT GGT ATA GTA CTT  317
                   ::  ::: ::  ::  ::  ::  ::  ::  ::  ::  ::  ::: ::  ::    :
HgAct(nt)     359  GTG CTC TCC CTG TAC GCT TCC GGC CGT ACC ACC GGT ATT GTC TTG  403
HgAct(aa)          Val Leu Ser Leu Tyr Ala Ser Gly Arg Thr Thr Gly Ile Leu Leu
ArAct302(aa)        -   -   -   -   -   -   -   -   -   -   -   -   -   -   -
```

```
ArAct302(nt) 318  GAT TCT GGA GAT GGC GTA TCT CAT ACC GTT CCC ATC TAT GAA GGT  362
                  ::  ::: ::  ::: ::: ::  ::  ::  ::  ::: ::  ::: ::  ::  ::
HgAct(nt)    404  GAC TCT GGT GAT GGC GTG TCA CAC ACT GTT CCT ATC TAC GAG GGA  448
HgAct(aa)         Asp Ser Gly Asp Gly Val Ser His Thr Val Pro Ile Tyr Glu Gly
ArAct302(aa)       -   -   -   -   -   -   -   -   -   -   -   -   -   -   -

ArAct302(nt) 363  TAT GCC CTC CCC CAT GCT ATT CTT CGT CTT GAT CTG GCT GGT CGT  407
                  ::  ::: ::  ::: ::: ::: ::  ::  ::: :::
HgAct(nt)    449  TAC GCC CTT CCC CAT GCT ATC CTG CGT CTG GAC TTG GCT GGA CGT  493
HgAct(aa)         Tyr Ala Leu Pro His Ala Ile Leu Arg Leu Asp Leu Ala Gly Arg
ArAct302(aa)       -   -   -   -   -   -   -   -   -   -   -   -   -   -   -

ArAct302(nt) 408  GAC CTT ACA GAC TAC CTG ATG AAG ATT CTT ACT GAA AGA GGC TAC  452
HgAct(nt)    494  GAC CTT ACT GAC TAC CTG ATG AAG ATC CTG ACT GAG CGT GGC TAC  538
HgAct(aa)         Asp Leu Thr Asp Tyr Leu Met Lys Ile Leu Thr Glu Arg Gly Tyr
ArAct302(aa)       -   -   -   -   -   -   -   -   -   -   -   -   -   -   -

ArAct302(nt) 453  ACT TTC ACT ACT ACA GCA GAA AGA GAA ATA GTT CGT GAT ATC AAA  497
HgAct(nt)    539  ACC TTC ACT ACC ACT GCT GAG CGA GAA ATC GTT CGT GAC ATT AAG  583
HgAct(aa)         Thr Phe Thr Thr Thr Ala Glu Arg Glu Ile Val Arg Asp Ile Lys
ArAct302(aa)       -   -   -   -   -   -   -   -   -   -   -   -   -   -   -

ArAct302(nt) 498  GAG AAG CTA TGC TAT GTA GCC CTT GAT TTT GAA CAA GAG ATG GCC  542
HgAct(nt)    584  GAA AAG TTG TGC TAT GTT GCC CTC GAC TTC GAG CAG GAA ATG ACC  628
HgAct(aa)         Glu Lys Leu Cys Tyr Val Ala Leu Asp Phe Glu Gln Glu Met Thr
ArAct302(aa)       -   -   -   -   -   -   -   -   -   -   -   -   -   -  Ala

ArAct302(nt) 543  ACA GCC GCA AGC TCA ACT TCT CTC GAG AAG AGT TAT GAG CTT CCT  587
HgAct(nt)    629  ACT GCT GCG TCG TCC TCC TCC CTA GAG AAG TCC TAC GAA CTT CCC  673
HgAct(aa)         Thr Ala Ala Ser Ser Ser Ser Leu Glu Lys Ser Tyr Glu Leu Pro
ArAct302(aa)       -   -   -   -   -  Thr  -   -   -   -   -   -   -   -   -

ArAct302(nt) 588  GAT GGA CAG ATT ATT ACC GGT AAT GAA CGA TTC CGT          626
HgAct(nt)    674  GAC GGT CAA GTT ATC ACC ATC GGT AAC GAG AGG TTC CGT TGT CCC  718
HgAct(aa)         Asp Gly Gln Val Ile Thr Ile Gly Asn Glu Arg Phe Arg
ArAct302(aa)       -   -   -  Ile  -   -   -   -   -   -   -   -   -

HgAct(nt)    719  GAG GCC CTC TTC GAA CC A   736
```

Fig. 2 (*cont.*)

majority of cases, these isoforms are encoded by unique genes which are regulated differentially during muscle development and are important in characterizing muscle fibre type (Bandman, 1992). An exception to this is found in *Drosophila*, where the MHC is encoded by a single copy gene containing 29 exons, which as a result of alternative RNA splicing potentially can produce up to 480 different isoforms (Bernstein *et al.*, 1993). So far, only 10 of the possible 480 MHC isoforms have been identified during development, and *Drosophila* MHC mutants are being used to assist in understanding the function of the various domains in

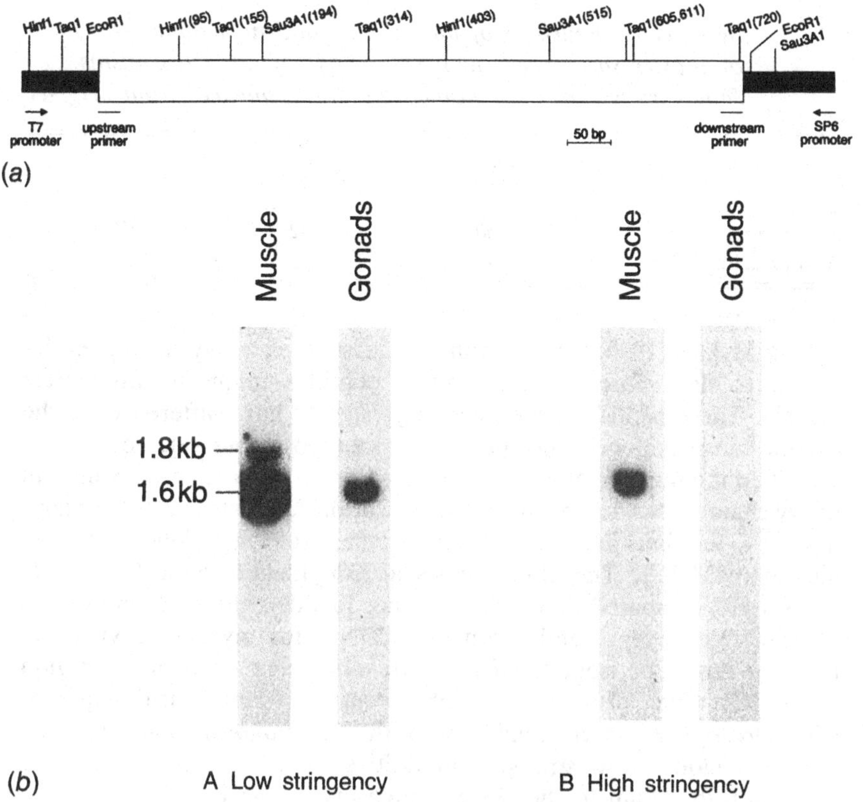

Fig. 3(*a*) Restriction map of the insert cloned in pHgAct (open bar) and part of the vector (closed bar). The coordinates (brackets) used to indicate the cleavage sites of the restriction endonucleases follow the numbering system shown in the nucleotide sequence of pHgAct in Fig. 2 (from Harrison & El Haj, 1994). (*b*) Northern blots of total RNA from *H. gammarus* muscle and gonads hybridized with HgAct. (A) Low stringency conditions in 2 × SSC, 0.1% SDS at 65 °C for 30 minutes (B) high stringency conditions in 0.1SSC, 0.1% SDS at 68 °C for 1 hour.

myosin. Cloning of myosin HC genes has taken advantage of the homologies or lack of homologies between different regions of the myosin molecule. The globular head region has been found to be well conserved with the ATPase active site and actin binding region, showing strong homologies between species. In contrast, the rod portion of the molecule is less well conserved and the 3′ UTR region is often used as an isoform determinant within species (Aigner & Pette, 1990).

Table 1. *Comparison of amino acid and nucleotide sequences for lobster sarcomeric actin with those of other organisms. Values are given as % homology with sequences from cHgAct*

	Insect	Vertebrate α actin	*Artemia*
Nucleotides	60	42	78
Amino acids	93	92	98

Li & Mykles (1990) have identified at least seven heavy chain peptides unique to slow muscle myosin and 11 peptides unique to fast muscle myosin. These peptides have been suggested to have differences in the flexible hinge region in the rod portion of myosin. As yet, no one has identified the number of genes encoding myosin in Crustacea. A myosin heavy chain cDNA has been cloned by Cotton & Mykles (1994) containing the *C*-terminus and the 3′ UTR of a fast isoform of lobster striated muscle (c4MHC). The clone shows a 73% identity with *Drosophila* myosin HC compared to a 49% sequence identity with chicken skeletal myosin. We have recently cloned a cDNA for myosin heavy chain from an Antarctic isopod, *Glyptonotus antarcticus* from the 5′ region (GaMHC) which also shows a 75% identity in amino acid sequence with *Drosophila*. Interestingly, both the *Glyptonotus* clone and the *Homarus* clone show stronger similarities to the mammalian cardiac myosins rather than to the skeletal myosins (Table 2).

The lobster c4MHC encodes a helical polypeptide with a 28-residue repeating pattern characteristic of the myosin rod in *Drosophila* with the last three amino acid residues non-helical, a typical feature of myosin HCs from mammalian striated muscle (Li & Mykles, 1990). The lobster clone and *Drosophila* clones recognize a mRNA of 6.6 kb which corresponds to the size of mRNAs in other species, although in *Drosophila* three different transcripts are present, 6.1, 6.6 and 7.1 kb. Our future studies include screening our genomic library for the isopod, *Glyptonotus antarcticus* using the cDNA from the head region of the myosin gene which will enable us to identify whether crustaceans like *Drosophila* have only one myosin gene which relies on alternate splicing to produce multiple isoforms.

Tropomyosin

Tropomyosin isoform diversity involves the expression of multiple genes, each of which codes for more than one isoform by the use of

Table 2. *Comparison of nucleotide identity of the isopod myosin heavy chain clone (MHC) with sequences of myosin heavy chain from other organisms*

Organism	Percentage identity	Overlap
Drosophila MHC	75.4	491
Atlantic bay scallop MHC *Aquipecten irradians*	62.3	547
Nematode *Coenarhabditis elegans*	63.7	559
Rabbit β Cardiac MHC	63.7	532
Rat β Cardiac MHC	61.3	530
Mouse α Cardiac MHC	61.1	529

alternatively spliced exons (Helfman, Ricci & Finn, 1988). The mechanism of alternate RNA splicing is quite common in a variety of species, e.g. *Drosophila* (Karlik & Fyrberg, 1986) and rat (Helfman *et al.*, 1988). Helfman *et al.* (1986) have shown that a single rat gene encodes both rat fibroblast TM1 and skeletal muscle B tropomyosin by alternate RNA splicing. In Crustacea, no evidence for alternate RNA splicing of tropomyosin has been found but a cDNA has been isolated by Cotton & Mykles (1994) for lobster of approximately 1.5 kb containing a 1 kb 3′ UTR and a coding sequence for the *C* terminal half of the protein. The tropomyosin clone has an 81% identify with *Drosophila* (Hanke & Storti, 1986). The tropomyosin clone recognizes two messages, 1.1 kb and 2.1 kb, but further information on this clone is awaiting publication. The myosin light chains have been characterized at the peptide level showing two isoforms of the light chain, but little is known as yet about the genes responsible (Li & Mykles, 1990).

Sarco/endoplasmic reticulum Ca-ATPase

Other genes have been cloned from crustacean muscle tissue; in particular, the enzymes responsible for the transport of calcium to the SR/ER after muscle contraction or cell activation, Ca-ATPase (SERCA). In vertebrates, there are three *SERCA* genes which are expressed in different muscle types: fast, slow and cardiac. Further determination of different enzyme isoforms is obtained by differential processing of the gene transcripts (Burk *et al.*, 1989). Isolation of clones from *Artemia* have highlighted the strong homologies between Arthropods. As with myosin, *Drosophila* has been shown to code for only one *SERCA* gene (Magyar & Varadi, 1990) and, in *Artemia franciscana*, only one genomic clone was isolated for the *SERCA* gene. Analysis of the genomic clones for the *SERCA* gene from *Artemia* has revealed that the gene is divided into 18 exons and is 65 kb long. The cDNA clones for *SERCA* recognize two mRNAs of sizes, 4.5 kb and 5.2 kb, which are produced by differential processing (Escalante & Sastre, 1993).

Regulatory hormones

cDNAs for some of the crustacean neurohormones have been cloned recently, and distribution of the mRNA has been localized to specific tissues. The neurohaemal organ, the sinus gland, produces CHH which is involved primarily in the regulation of blood sugar levels and glycogen metabolism (Webster & Keller, 1988). cDNAs for CHH have been cloned for the crayfish, *Orconectes*, sp. (De Kleijn *et al.*, 1994), the lobster, *Homarus* sp. (Tensen, De Kleijn & Van Herp, 1991; Rotllant *et al.*, 1993) and a common shore crab, *Carcinus maenas* (Weidermann *et al.*, 1989). There are strong sequence homologies for all these species. De Kleijn *et al.* (1994) suggests that the existence of two isoforms of the CHH neuropeptide is due to post-translational modification of identical CHH species originating from two genes. A partial DNA sequence for MIH from *Carcinus maenas* and the white shrimp, *Panaeus vannamei* has been cloned (Klein *et al.*, 1993; Sun, 1994) and the MIH-like mRNA has a molecular size of 2.0 kb. CHH and MIH may be very closely related. Other regulatory hormones such as vitellogenesis inhibiting hormone (VIH) or gonad inhibiting hormone (GIH) which control reproduction by delaying secondary vitellogenesis and gonad maturation have been sequenced (Soyez *et al.*, 1991). Oligos derived from protein sequences and PCR-generated clones have been used for *in situ* hybridization to localize mRNA production to secretory cells in the eyestalks of *Homarus americanus* (Laverdure *et al.*, 1992; Rotllant *et al.*, 1993). Little data on the regulatory sequences from the genomic clones are available on these hormones.

Other genes involved in growth

Increasingly, more genes are being cloned from *Artemia*, for example, those involved during development such as *engrailed* homeobox (Manzanares, Marco & Garesse, 1993) and Na^+/K^+-ATPase (Macias,

Palmero & Sastre, 1991). The strong homologies between genes cloned for insects and those for Crustacea are being found in all the genes isolated to date which is of interest from an evolutionary viewpoint. The *engrailed*-like gene of *Artemia* has four domains which are extremely well conserved within Arthropods (Manzanares *et al.*, 1993).

Crustacean promoter regulatory sequences

Little information is available on the promoters and regulatory regions of crustacean genes. The majority of studies have focused on isolating cDNA clones for *in vivo* tissue expression studies rather than for genomic sequence analysis. Some work has been carried out on the *SERCA* gene from *Artemia*, with primer extension and nuclease S1 protection experiments used to characterize the 5′ end of the gene (Escalante & Sastre, 1994). Two origins of transcription have been located, separated by 30 bp, which have been found previously in genes without TATA boxes in the gene promoter. However, in *Artemia SERCA* gene, a TATA box has been identified and the two initiation sites may be involved in RNA splicing to produce differentially sized transcripts. Other regulatory sequences identified include a CCAAt box, TCF-1, E2A and H4TF-1 binding sites. In addition, there are muscle-specific regulatory regions, eight CANNTG motifs that are possible binding sites of the MyoD family, MEF-2 sites and a CArG element (Escalante & Sastre, 1994), although no functional analysis to confirm the identity of this region has been carried out as yet. Further work is needed on the genomic 5′ regions of the sarcomeric protein genes described above to understand transcriptional regulation of muscle growth and development.

Growth of crustacean muscle

Longitudinal and cross-sectional hypertrophy of the muscle occurs during the 1–2 week period during the lead up to, and following, the moult in crabs and lobsters (El Haj *et al.*, 1984; Houlihan & El Haj, 1985). In the walking leg muscle of *Carcinus maenas* and *Homarus americanus*, sarcomere addition is at the end of the fibres which are attached to the new exoskeleton (El Haj, Harrison & Chang, 1994; Houlihan & El Haj, 1985), and myofibres increase in width by addition to the number of myofibrils and myofibrillar splitting as with vertebrate muscle (El Haj *et al.*, 1984). Total protein synthesis rates in the muscle are elevated during the pre-moult and post-moult period in *Carcinus* (Houlihan & El Haj, 1987) and *Homarus* sp. (El Haj *et al.*, in press) to account for this growth. In the isopod, which shows biphasic moulting, muscle protein synthesis rates differ between the anterior and posterior portions of the ventral longitudinal muscle of the animal corresponding to when growth is occurring (Whiteley *et al.*, 1995).

Different muscle groups in crustacea may grow at different times through the moult cycle by regulating rates of degradation through proteinases. This is the case in the claw muscles which undergo atrophy during pre-moult to allow the muscle to be pulled through the basiischium joint during ecdysis. The most extreme example of this is the Bermudan land crab, *Geocarcinus lateralis* which atrophies by approx. 30–60% during the pre-moult phase (Skinner, 1966). Mykles & Skinner (1985) have demonstrated that there are two peaks in protein synthesis in the pre-moult and post-moult phase in the claw muscle. Our results have shown, in the lobster, that muscle protein synthesis rates are elevated in all muscle groups, both leg, abdomen and claw, during the pre-moult (El Haj *et al.*, in press). Thus claw muscle is atrophying when leg and abdominal muscle is growing. To account for this, there must be a large tissue-specific increase in enzyme activity in the claw muscle during this stage. Mykles & Skinner (for review, see 1990) have identified the role of calcium-dependent proteinases and multicatalytic proteinases in degrading the myofibrillar proteins with preferential degradation of thin filament proteins such as actin and tropomyosin versus thick filaments such as myosin. One theory proposed is that degradation of the old skeleton during pre-ecdysis leads to elevation in calcium levels in the tissues which may, in some way, trigger elevations in proteinase activity (Mykles & Skinner, 1990). This would allow for selective activation in certain tissues.

Levels of expression of sarcomeric proteins during the moult cycle

Actin mRNA levels have been measured over the moult cycle in a number of crustacean species, and differing patterns are emerging. In *Carcinus*, early studies using heterologous probes for mammalian actin showed elevated mRNA levels in the pre-moult and post-moult phase (Whiteley, Taylor & El Haj, 1992). A similar pattern was seen in the crayfish, *Austropotamobius pallipes* using an *Artemia* actin cDNA to probe actin mRNA levels over the moult cycle in both claw and leg muscle with elevated levels in the pre- and post-moult periods compared to the inter-moult (El Haj *et al.*, 1992). In both these studies, the animals were selected from wild populations and moult staged according to external characteristics. Recently, our studies on a laboratory population of the American lobster, *Homarus americanus*, held at 20 °C has shown that actin mRNA levels remain constant when measured at daily intervals throughout the moult cycle in leg abdomen and claw muscle when actin protein levels are increasing. Fig. 4(*a*) shows the constant levels of actin mRNA through the moult in the claw muscle using the *Homarus* sp. cDNA, HgAct. In Fig. 4(*b*) the claw muscle,

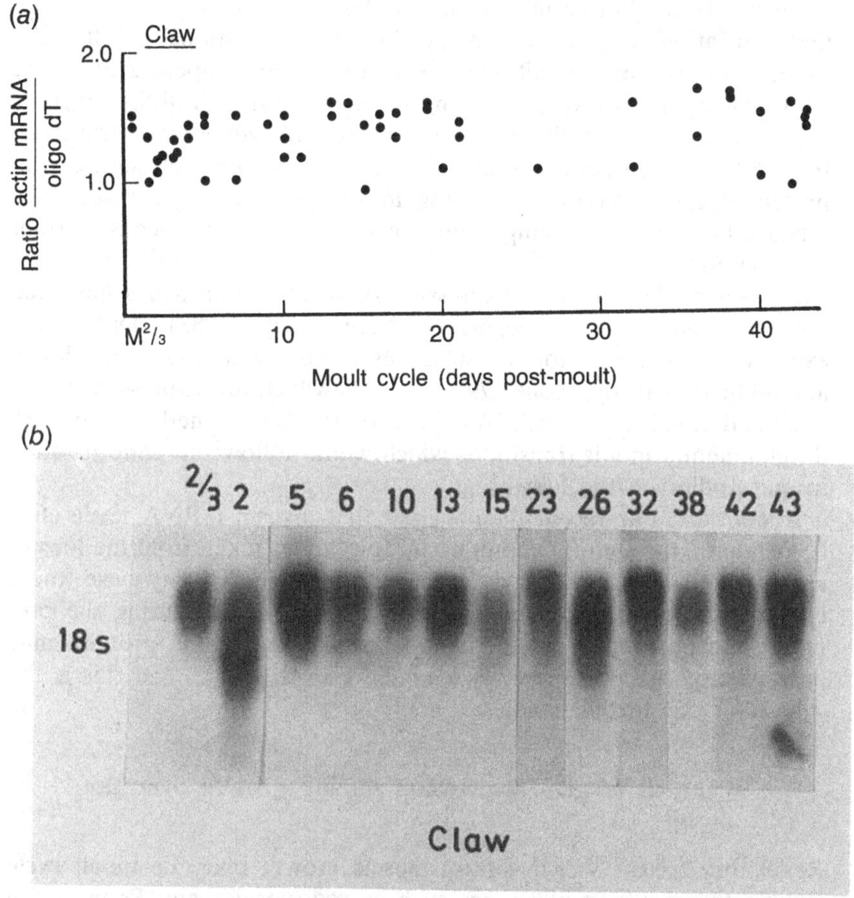

Fig. 4(*a*) Relationship between actin mRNA levels/ oligo dT and days through the moult cycle in the claw muscle of the American lobster, *Homarus americanus*. The *x* axis indicates the days post-moult and the animals moult cycle consisted of a 43–45 day cycle. (*b*) Northern blot of total RNA extracted from the claw muscle of the American lobster at different stages of the moult cycle hybridized with an 18S ribosomal cDNA. The numbering indicates the days post-moult as above (from El Haj *et al.*, in press).

hybridized with a ribosomal 18S cDNA, demonstrates constant levels of ribosomal mRNA through the moult, indicating that elevations in protein synthesis are not due to increased ribosomal content. In the lobster it may be that actin protein production is regulated at the translational level by ribosomal activity or post-translational conversions between pools of G/F actin proteins during the moult cycle. In addition, there may be a relationship between holding temperatures and transcription/translation rates which is mapped over the traditional moult cycle. In our studies, intermoult isopods held at two temperatures 4 and 14 °C, have shown a difference in Q_{10} of 1.9 in actin mRNA levels in the muscle. However, the elevations in protein synthesis and transcription at higher temperatures may also be accompanied by increases in protein degradation rates resulting in no net growth of tissue. The relationship between temperature and moult stage needs further investigation.

In insects, the temporal expression of the six actin genes has been measured during development (Fyrberg *et al.*, 1983). Differential expression was found for the six genes examined during pupal, larval and adult life. In the clones 79B and 88F, which are expressed during adult and pupal stages, mRNA expression levels remained high through developmental moult transitions which would follow the pattern found in our studies on the lobster.

In contrast, our recent data suggest that myosin mRNA levels may be regulated through the moult cycle. In samples taken from the lobster leg muscle held at 20 °C at different days through a 45-day cycle, there is evidence for some fluctuation with elevated levels during the pre-moult and post-moult phase. It appears that sarcomeric proteins may be regulated in different ways over the growth phase, and this is the subject of our further studies.

Control of gene expression during growth over the moult

Regulatory factors which control muscle growth over the moult cycle include the moulting hormones such as ecdysteroids and Ponasterone A but also influences such as passive stretch. Stretch is imposed on crustacean leg muscles immediately during ecdysis when the new exoskeleton is expanding prior to calcification. Passive stretch has been shown to promote hypertrophy in vertebrate muscle (Goldspink, 1980), and effects on sarcomeric protein gene expression have been demonstrated. In Crustacea, Houlihan & El Haj (1985) have shown that experimentally induced stretch results in longitudinal growth in the myofibrils varying in specific regions between 5 and 34% of the initial length, of the carpopodite extensor muscles of the intermoult crab, *Carcinus maenas*. A time course of the response measured in the lobster *Homarus gammarus* revealed that longitudinal growth of the myofibrils occurred after 1 week of stretch with the major growth phase in week 2. Corresponding to the growth phase at week 2 was a 70% elevation in actin mRNA abundance. The time course of this response, however, is slow compared to the rapid growth phase over ecdysis. It has been suggested that the passive stretch imposed in these experiments is much less than that which the leg muscle experiences during cuticular stretch. The greater amounts of stretch imposed during ecdysis may result in a more rapid response.

Ecdysteroids fluctuate through the moult cycle with peaks at pre- and post-moult corresponding to the growth phase (Snyder & Chang, 1991). Hormonal influences have been investigated in relation to the formation of the new exoskeleton prior to ecdysis. Ecdysteroids have been shown to have profound effects on integumentary tissues *in vitro* with a range of effects observed (Soumoff & Skinner, 1988; Paulson & Skinner, 1991). Skinner, Kumari & O'Brien (1992) suggest that the large variation in responses recorded may reflect the way in which changing 20HE concentrations with time through the moult control the expression of cascades of exoskeletal proteins. Crab muscle tissue has been demonstrated to be ecdysteroid responsive (Whiteley *et al.*, 1992). This response may be species specific, as lobster muscle has not been found to be ecdysteroid responsive *in vitro* (El Haj *et al.*, in press). A clear picture of the actions of ecdysteroids on muscle tissue is yet to emerge. Injections of pre-moult concentrations of 20 hydroxyecdysone *in vivo* results in elevated protein synthesis in the muscle up to 3 days after injection. This response is seen in leg abdomen and claw muscle (El Haj *et al.*, in press). El Haj *et al.* (1994) have shown ecdysteroid receptor immunoreactivity in the walking leg extensor muscle of the American lobster using a monoclonal antibody to EcR from *Drosophila*. Western blots using the EcR antibody hybridized with a 95–110 kD protein from the lobster tissues. If ecdysteroid receptors are present in the muscle, direct effects should be apparent; however, these may be on intermediary genes rather than directly on the sarcomeric protein genes. In insects, a number of ecdysteroid responsive genes have been isolated, and the cascade of responses partially mapped out (Hurban & Thummel, 1993). Lobster homologues have yet to be isolated, and this is the subject of our further work.

Intermediary factors may also play a role in mediating hormonal responses *in vivo*, which are not present *in vitro*. Various growth factors have been proposed to play a role as mediators of growth although as yet none have been characterized (Skinner *et al.*, 1992). Preliminary studies in the lobster have shown that insulin-like factors are present (Sanders, 1983). Our studies using a modified mammalian radioimmunoassay using IGF antibodies made from recombinant sequences have recorded small concentrations of IGFs in green gland and muscle, and mRNA species have been identified using heterologous probes. However, homologous probes are necessary before detailed studied can be carried out to investigate the potential role of growth factors in muscle growth in Crustacea.

References

Aigner, S. & Pette, D. (1990). *In situ* hybridization of slow myosin heavy chain mRNA in normal and transforming rabbit muscles with the use of a nonradioactively labelled cRNA. *Histochemistry*, **95**, 11–18.

Bandman, E. (1992). Contractile protein isoforms in muscle development. *Developmental Biology*, **154**, 273–83.

Bennet, D.B. (1974). Growth of the edible crab (*Cancer pagurus*) L. off Southwest England. *Journal of Marine Biological Association*, **54**, 803–23.

Bernstein, S.I., O'Donnell, P.T. & Cripps, R.M. (1993). Molecular genetic analysis of muscle development structure and function in *Drosophila. International Review of Cytology*, **143**, 63–146.

Buckingham, M.E. & Minty, A. (1983). Contractile protein genes. In *Eukaryotic Genes: Their Structure, Activity and Regulation*, ed. N. Maclean, S. Gregory & R. Flavell, pp. 35–45. London: Butterworth.

Burk, S.E., Lytton, J., Maclennan, D.H. & Shull, G.E. (1989). cDNA cloning, functional expression and mRNA tissue distribution of a 3rd organellar Ca^{2+} pump. *Journal of Biological Chemistry*, **264**, 18561–8.

Chang, E.S. (1993). Comparative endocrinology of moulting and reproduction; insects and crustaceans. *Annual Review Entomology*, **38**, 161–80.

Cotton, J.L.S. & Mykles, D.L. (1994). Cloning of a crustacean myosin heavy chain isoform: exclusive expression in fast muscle. *Journal of Experimental Zoology*, **267**, 578–86.

Crothers, J.H. (1967). The biology of the shore crab, *Carcinus maenas* (L.) *Field Study*, **2**, 407–34.

De Kleijn, D.P.V., Janssen, K.P.C., Martens, G.J.M. & Van Herp, F. (1994). Cloning and expression of two crustacean hyperglycemic hormone mRNAs in the eyestalk of the crayfish *Orconectes limosus. European Journal of Biochemistry*, **224**, 623–9.

El Haj, A.J., Clarke, S., Harrison, P. & Chang, E. (In press). *In vivo* muscle protein synthesis rates in the American lobster, *Homarus armericanus*, during the moult cycle and in response to 20-hydroxyecdysone. *Journal of Experimental Biology*.

El Haj, A.J., Harrison, P. & Chang, E.S. (1994). Localization of ecdysteroid receptor immunoreactivity in eyestalk and muscle tissue of the American lobster, *Homarus americanus*. *Journal of Experimental Zoology*, **270**, 343–9.

El Haj, A.J., Harrison, P. & Whiteley, N.M. (1992). Regulation of muscle gene expression in crustacea over the moult cycle. In *Molecular Biology of Muscle*, ed. A.J. El Haj, pp. 151–165. Cambridge: Cambridge University Press.

El Haj, A.J. & Houlihan, D.F. (1987). *In vitro* and *in vivo* protein synthesis rates in a crustacean muscle during the moult cycle. *Journal of Experimental Biology*, **127**, 413–26.

El Haj, A.J., Govind, C.K. & Houlihan, D.F. (1984). Growth of lobster leg muscle fibres over intermoult and moult. *Journal of Crustacean Biology*, **4**, 536–45.

Escalante, R. & Sastre, L. (1993). Similar alternative splicing events generate two sarcoplasmic or endoplasmic reticulum Ca ATPase isoforms in the crustacean *Artemia franciscana* and in vertebrates. *Journal of Biological Chemistry*, **268**, 14090–5.

Escalante, R. & Sastre, L. (1994). Structure of *Artemia franciscana* sarco/endoplasmic reticulum Ca ATPase gene. *Journal of Biological Chemistry*, **269**, 13005–12.

Fyrberg, E.A., Bond, B.J., Hershey, N.D., Mixter, K.S. & Davidson, N. (1980). The actin genes of *Drosophila*: a dispersed multigene family. *Cell*, **19**, 365–78.

Fyrberg, E.A., Bond, B.J., Hershey, N.D., Mixter, K.S. & Davidson, N. (1981). The actin genes of *Drosophila*: protein coding regions are highly conserved but intron positions are not. *Cell*, **24**, 107–16.

Fyrberg, E.A., Mahaffey, J.W., Bond, B.J. & Davidson, N. (1983). Transcripts of six *Drosophila* actin genes accumulate in a stage and tissue specific manner. *Cell*, **33**, 115–23.

Goldspink, G. (1980). Growth of muscle. In *Development and Specialization of Skeletal Muscle*, ed. G. Goldspink, SEB Seminar Series 7, Cambridge: Cambridge University Press.

Hanke, P.D. & Storti, R.V. (1986). Nucleotide sequence of a cDNA clone encoding a *Drosophila* muscle tropomyosin II isoform. *Gene*, **45**, 211–14.

Harrison, P. & El Haj, A.J. (1994). Actin mRNA levels and myofibrillar growth in leg muscles of the European lobster (*Homarus gammarus*) in response to passive stretch. *Molecular Marine Biology and Biotechnology*, **3**, 35–41.

Hartnoll, R.G. (1982). Growth. In *The Biology of Crustacea, vol. 2*, ed. L.G. Abele, NY: Academic Press.

Helfman, D.M., Cheley, S., Kuusmanen, E., Finn, L.A. & Yamawaki-Kataoka, Y. (1986). Nonmuscle and muscle tropomyosin isoforms are expressed from a single gene by alternate RNA splicing and polyadenylation. *Molecular Cell Biology*, **6**, 3582–95.

Helfman, D.M., Ricci, W.M. & Finn, L.A. (1988). Alternative splicing of tropomyosin premRNAs *in vitro* and *in vivo*. *Genes and Development*, **2**, 1627–38.

Houlihan, D.F. & El Haj, A.J. (1985). An analysis of muscle growth. In *Factors in Adult Growth*, ed. A. Wenner. Amsterdam: Balkema Press.

Houlihan, D.F. & El Haj, A.J. (1987). Muscle growth. In *Crustacean Growth*, ed. A. Wenner, pp. 15–30. Rotterdam: Balkema Press.

Hurban, P. & Thummel, C.S. (1993). Isolation and characterization of 15 ecdysone inducible *Drosophila* genes reveal unexpected complexities in ecdysone regulation. *Molecular and Cellular Biology*, **13**, 7101–11.

Karlik, C. & Fyrberg, E. (1986). Two *Drosophila* tropomyosin genes: structural and functional aspects. *Molecular Cell Biology*, **6**, 1965–73.

Klein, J.M., de Kleijn, D.P.V., Hunemeyer, G., Keller, R. & Weidermann, W. (1993). Demonstration of the cellular expression of genes encoding molt inhibiting hormone and crustacean hyperglycaemic hormone in the eyestalk of the shore crab, *Carcinus maenas*. *Cell and Tissue Research*, **274**, 515–19.

Laverdure, A.M., Bruezet, M., Soyez, D. & Becker, J. (1992). Detection of mRNA encoding Vitellogenesis inhibiting hormone in neurosecretory cells of the x-organ of *Homarus americanus* by *in situ* hybridization. *General and Comparative Endocrinology*, **87**, 443–50.

Li, Y. & Mykles, D.L. (1990). Analysis of myosins from lobster muscles: fast and slow isozymes differ in heavy chain composition. *Journal of Experimental Zoology*, **255**, 163–70.

Macias, M.-T. & Sastre, L. (1990). Molecular cloning and expression of four actin isoforms during *Artemia* development. *Nucleic Acids Research*, **18**, 5219–25.

Macias, M.-T., Palmero, I. & Sastre, L. (1991). Cloning of a cDNA encoding an *Artemia franciscana*, Na/K ATPase alpha subunit. *Gene*, **105**, 197–204.

Magyar, A. & Varadi, A. (1990). Molecular cloning and chromosomal localization of a sarco/endoplasmic reticulum Ca^{2+} ATPase of *Drosophila*. *Biochemical and Biophysical Research Communications*, **173**, 872–7.

Manseau, L.J. Ganetsky, B. & Craig, E.A. (1988). Molecular and genetic characterisation of the *Drosophila* 87E actin gene region. *Genetics*, **119**, 407–20.

Manzanares, M., Marco, R. & Garesse, R. (1993). Genomic organization and developmental pattern of expression of the engrailed gene from the brine shrimp *Artemia*. *Development*, **118**, 1209–19.

Minty, A.J., Caravatti, M., Robert, B., Cohen, A., Daubas, P., Weydert, A., Gros, F. & Buckingham M.E. (1981). Mouse actin mRNAs. *Journal of Biological Chemistry*, **256**, 1008–14.

Mykles, D.L. & Skinner, D.M. (1990). Atrophy of crustacean somatic muscle and the proteinases that do the job. A review. *Journal of Crustacean Biology*, **10**, 577–94.

Mykles, D.L. & Skinner, D.M. (1985). Muscle atrophy and restoration during molting. In *Crustacean Issues, Vol. II, Crustacean Growth*, ed. A.M. Wenner, pp. 31–46. Amsterdam: Balkema Press.

Ortega, M.-A., Marcias, M.-T., Martinez, J.L., Palmero, I. & Sastre, L. (1992). Expression of actin isoforms in *Artemia*. In *Molecular Biology of Muscle*. ed. A.J. El Haj, pp. 131–137. Cambridge: Cambridge University Press.

Paulson, C.R. & Skinner, D.M. (1991). Effects of 20 hydroxyecdysone on protein synthesis in tissues of the land crab *Gecarcinus lateralis*. *Journal of Experimental Zoology*, **257**, 70–9.

Rottlant, G., de Kleijn, D., Charmantier Daures, M., Charmantier, G. & Van Herp, F. (1993). Localization of CHH and GIH in the eyestalk of *Homarus gammarus* larvae by immunocytochemistry and *in situ* hybridization. *Cell and Tissue Research*, **271**, 507–12.

Sanders, B. (1983). Insulin like peptides in the lobster *H. americanus* 1. Insulin immunoreactivity. *General Comparative Endocrinology*, **50**, 366–73.

Skinner, D.M. (1966). Breakdown and reformation of somatic muscle during the molt cycle of the land crab, *Gecarcinus lateralis*. *Journal of Experimental Zoology*, **163**, 115–24.

Skinner, D.M., Kumari, S.S. & O'Brien, J.J. (1992). Proteins of the crustacean exoskeleton. *American Zoologist*, **32**, 470–84.

Snyder, M.J. & Chang, E.S. (1991). Ecdysteroids in relation to the moult cycle of the American lobster, *H. americanus* 1. Haemolymph titres and metabolites. *General Comparative Endocrinology*, **65** 469–77.

Soumoff, C. & Skinner, D.M. (1988). Ecdysone 20 mono-oxygenase activity in land crabs. *Comparative Biochemistry and Physiology*, **91C**, 139–44.

Soyez, D., La Caer, J.P., Noel, P.Y. & Rosier, J. (1991). Primary structure of two isoforms of the VIH from the lobster, *Homarus americanus*. *Neuropeptides*, **20**, 25–32.

Sparrow, J.C., Drummond, D.R., Hennessey, E.S., Clayton, J.D. & Lindegaard, F.B. (1992). *Drosophila* actin mutants and the study of myofibrillar assembly and function. In *Molecular Biology of Muscle*, ed. A.J. El Haj, pp. 111–137. Cambridge: Cambridge University Press.

Sun, P.S. (1994). Molecular cloning and sequence analysis of a cDNA encoding molt-inhibiting hormone-like neuropeptide from the white shrimp, *Panaeus vannamei. Molecular Marine Biology and Biotechnology*, **3**, 1–6.

Tensen, C.P., De Kleijn, D.P.V. & Van Herp, F. (1991). Cloning and sequence analysis of a cDNA encoding two crustacean hyperglycemic hormones from the lobster, *Homarus americanus. European Journal of Biochemistry*, **200**, 103–6.

Webster, S.G. & Keller, R. (1988). Physiology and biochemistry of crustacean neurohormonal peptides. In *Neurohormones in Invertebrates*, ed. M.C. Thorndyke and G.J. Goldsworthy. Society Experimental Biology Seminar Series 33, pp. 173–96, Cambridge: CUP.

Weidermann, W., Gromoll, J. & Keller, R. (1989). Cloning and sequence analysis of cDNA for precursor of a CHH. *FEBS Letters*, **257**, 31–4.

Whiteley, N.M., Taylor, S.E., Chang, E.S. & El Haj, A.J. (1995). Muscle growth in an isopod crustacean over the biphasic moult. *Physiological Zoology*, **68(4)**, 148.

Whiteley, N.M., Taylor, E.W. & El Haj, A.J. (1992). Actin gene expression during muscle growth in *Carcinus maenas. Journal of Experimental Biology*, **167**, 277–84.

Q. XU, K. GRIFFIN, R. PATIENT and N. HOLDER

Use of the zebrafish for studies of genes involved in the control of development

The zebrafish as a model system for the study of vertebrate development

Over the past ten years the zebrafish has emerged as a key model system for the study of vertebrate development. This has occurred primarily because of the promise of the system for developmental genetic studies, but, in addition to the necessary features of an animal which can be used for genetics, there are a range of experimental approaches which have proved successful in studies of tissue interactions and gene function. Such methods include cell transplantation and the analysis of gene function by injection of RNA or antibodies into the fertilized egg. The object of this review is to outline the main features of zebrafish development and the methods which have been used in order to identify and study the role of developmentally important genes.

It is important to point out that gene function analyses in the zebrafish, whether they be by mutational screens or RNA injection, are carried out against an increasing and extensive knowledge of basic embryology. Thus the transparency and rapid development of the embryo has been exploited to great effect by Kimmel and his colleagues in establishing a basic fate map (Kimmel, Warga & Schilling, 1990) using cell marking experiments in which flourescent dyes are injected into single cells. This kind of analysis has also led to a framework understanding of the lineage relationships of cells in the blastula and an assessment of the timing of cell fate lineage restrictions. It is now clear, despite some data to the contrary (Strehlow & Gilbert, 1993; Wilson, Helde & Grunwald, 1993) that there is no clear restriction to cell fate in the zebrafish until gastrulation begins (Kimmel & Warga, 1986, 1988; Helde *et al.*, 1994). Consistent with this is the demonstration that the dorso-ventral axis, despite being established during blastula stages, occurs randomly with respect to the initial blastomere divisions (Abdelilah *et al.*, 1994).

The zebrafish owes its elevation from common pet shop aquarium fish to one of the few model systems for the study of vertebrate development largely to the 'Oregon School'. George Streisinger, at the University of Oregon at Eugene, first settled on the zebrafish for his pioneering genetic studies. He chose the zebrafish over the Japanese medaka largely because the eggs of the zebrafish are freely dispersed, whereas the medaka female keeps her eggs glued together close to the cloaca, making it difficult to search for mutants affecting early development. The medaka is still used for developmental studies, and has been used recently for the analysis of mesoderm formation (Wittbrodt & Rosa, 1994).

Developmental genetics in the zebrafish

Streisinger established a number of methods for studying zebrafish including the generation of isogenic homozygous diploid lines and the screening of haploid embryos for developmentally interesting mutations (Kimmel, 1989). The zebrafish develops for several days as a haploid and eventually dies; a screening programme based on haploid embryos has the advantage that recessive mutations present in the female can be revealed in a single generation. This method has been used effectively by Charles Kimmel's laboratory at the University of Oregon to reveal several mutants either induced or present in the genetic background of the zebrafish stock held in Eugene. Such mutants include *spadetail* (Kimmel *et al.*, 1989), *cyclops* (Hatta *et al.*, 1991) and *no-tail* (Schulte-Merker *et al.*, 1994); lines which have subsequently been extensively studied, the last being important for studies of patterning of the axial midline of the embryonic axis.

Since the first attempts to establish zebrafish for developmental genetics, the single most important step has been the selection of this system for a number of major screens using chemical mutagenesis and analysis of sib-crosses in the F2 generation to reveal mutations originally induced into male germ cells by chemical mutagenesis using ENU. The strategy, methodology and introductory results from these screens have recently been published by two of these laboratories (Mullins *et al.*, 1994; Solnica-Krezel, Schier & Driever, 1994; Driever *et al.*, 1994), and this material will not be repeated here.

Together with the continuing screening in Oregon, these two large-scale screens are now at the stage where mutations are being characterized genetically and morphologically. It is clear that the screens have revealed many mutants of developmental interest affecting the major organ systems and structures of the embryo.

From mutation to gene

The identification of a gene by mutation is only part of the process of functional analysis. The phenotype gives a clue as to the function of the gene, but a complete analysis of its role awaits its molecular characterization. At present, this is a difficult step in the zebrafish. There are a number of reasons for this, and the position is changing all the time as new methods are being worked out. We will focus on one method, insertional mutagenesis, that is based upon the injection of DNA into the fertilized egg which inserts into the host genome. If this DNA falls within an active region of chromatin, it may cause a mutation.

Before attempting this, we wished to examine the percentage of transmission through the germ line of DNA incorporated following injection into the fertilized embryo at the 1–2 cell stage (see Table 1). Such studies have been tried before in the zebrafish, and the results we report below are consistent with the first reports of successful transgenesis (Stuart *et al.*, 1990; Culp, Nusslein-Volhard & Hopkins, 1991). In our studies the *lacZ* gene under the control of a strong promoter, that of the cytomegalovirus/thymidine kinase (CMVtk) gene, was injected. In these cases after injection, approximately 10% of cells express β-gal by the onset of gastrulation and few differentiated cells are expressing by 24 hours of development. Thus DNA injected into the zebrafish egg appears to distribute to a low percentage of blastomeres and, most likely, are expressed in even fewer cells. The likelihood of such DNA ending up in germ cells in the blastula is therefore slim, and the percentage of transmission low.

In our preliminary insertional mutagenesis experiments, a gene trap construct was injected into the fertilized egg (Gossler *et al.*, 1989).

Table 1 *Efficiency of transgene expression in zebrafish embryos*

Constructs	*LacZ*	RNA	Estimate of cells/embryo
CMVtk.lacZ	82%	–	5–10%
Muscle actin/lacZ	63%	–	1–2% muscle
pGT1.8K	3%	–	0.01%
	–	43%	0.3%

(pGT1.8K gives 5% germ line transmission)

The construct was 1.8 kb of DNA encoding the mouse engrailed two genomic DNA with flanking regions for intron/exon splice acceptor sequences fused with *lacZ*. The idea of this gene trap, originally designed and used successfully by Gossler *et al.* (1989) in ES cells and mouse embryos, is that, if the injected DNA is inserted into the splice site, endogenous gene expression will be mimicked by *lacZ* expression. This will allow the assessment of expression, and therefore, the likely generation of a mutation due to insertion by revealing functional β-gal enzyme with a colour reaction. Potentially this can be done either in live embryos in the injected generation or in F2 or F3 generations in fixed embryos. In experiments in which 60 families originally injected with the gene trap were bred through to the F2, only five embryos (6%) contained the injected DNA as assessed by PCR using primers designed to amplify part of the β-gal sequence.

This mosaic behaviour of injected DNA is also a limitation for experiments aimed at promoter analysis. The design of such experiments is generally to fuse the promoter sequence of interest to a reporter gene, such as *lacZ*, inject this into the egg and examine expression relative to the temporal and spatial expression of the gene normally under the control of the promoter. Such analysis can be performed in the zebrafish but, due to mosaicism, the conclusion must be drawn from the analysis of a large number of embryos (Reinhard *et al.*, 1994).

Considerable progress has been made in generating a genetic recombination map for genes relative to the chromosomes. Methods based on PCR (the so-called RAPD method) have been applied by John Postlethwait's laboratory (Postlethwait *et al.*, 1994) in conjunction with the generation of haploids and polymorphic short sequences to produce a map with, as from September 1994, over 400 loci. Genes newly isolated by mutation, and genes cloned by homology to those in other species, are now being added to the map. It is clear that the map will have many markers within a short time, and will be a valuable resource for molecular identification of novel mutations either by mapping of cloned genes to the same site as an identified mutation or by positional cloning, although the latter has yet to be achieved in the zebrafish.

The analysis of gene function by RNA and antibody injection

Antibodies, particularly those that have been shown to have functional blocking activity, may be used as a tool for selectively blocking function following injection into the egg. We have achieved a functional knockout of at least part of the expression domain of the DNA binding nuclear regulatory protein *Pax 2* using this approach (Krauss *et al.*, 1992). In the zebrafish, *Pax 2* normally expresses in the early neural plate in the prospective midbrain/hindbrain boundary region and subsequently in the eye stalks and in various spinal neurons. If antibody to Pax-2 is injected into the fertilized egg, the caudal midbrain region that normally forms the cerebellum fails to form implicating this transcription factor as a key regulator of this region. Despite this clear example of the success of this strategy, we are not aware of other published cases of functional interuption by antibodies in the zebrafish.

Injection of RNA into the zebrafish egg or early blastomeres has proved a much more reliable method for analysis of gene function. In our laboratory we have used synthetic RNA encoding a number of proteins, cloned into the vector pSP64t (Kreig & Melton, 1984). This vector generates RNA flanked by globin sequences from the 3′ untranslated region. These sequences are thought to stabilize the RNA of interest, and do not interfere with the protein following translation. Following the injection of a number of different RNAs, it is evident that this approach can be used for overexpression of the wild-type protein and for dominant negative or dominant positive type experiments to analyse function.

Injected RNA distributes mosaicly in the zebrafish, but it is much more broadly distributed than injected DNA. Following *in situ* hybridization to the untranslated globin sequences or the coding sequences of the species of interest, we have followed the distribution of several separate injected messages and in both cases the RNA distributed to between 25 and 50% of blastomeres (Fig. 1). We have also examined the survival of the RNA and in two cases (Sek: Nieto *et al.*, 1992 and zebrafish *Hoxa-1*: Alexandre *et al.*, 1996) and it has been shown to be present by *in situ* hybridization in both cases throughout gastrulation and early neurulation stages but to have disappeared by tail bud. However, the stability of injected synthetic RNAs is variable. For example, *Pax6* RNA survives for about 4 hours after injection and the protein is rapidly turned over (Macdonald & Wilson, pers. comm.); zebrafish *Gata 3* RNA (Neave *et al.*, 1995) also survives about 4 hours.

We have used RNA injection to study proteins involved in various phases of the embryonic patterning process. These include intercellular signalling proteins (FGF–Griffin, Patient & Holder, 1995; Hedgehog–Macdonald *et al.*, 1995), membrane receptors (Sek, Xu *et al.*, 1995; FGFR–Griffin *et al.*, 1995–and dominant negative forms of these) and nuclear proteins (*Hoxa-1* and *Gata 1 and 3*; Alexandre *et al.*, 1996;

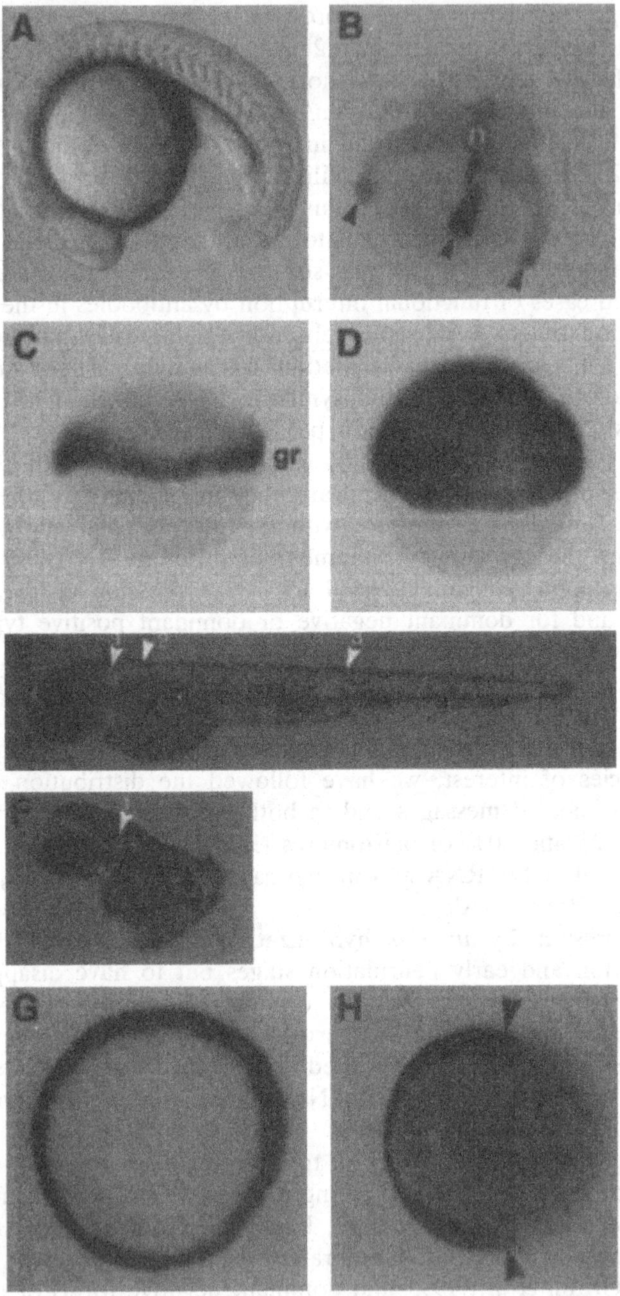
A
B
C
gr
D
F
G
H

Neave, Patient & Holder, unpublished results). In each case it has been important to titrate the amount of RNA that is injected to be clear about levels of toxicity and variations in phenotype generated as a result of variations in protein concentration. Toxicity and non-specific effects seems particularly a concern with nuclear proteins. It is also important to think carefully about controls relevant for a particular experiment. We standardly inject frameshift, antisense sequences plus other related members of families.

Tracking the distribution of the RNA may also be important in terms of understanding the eventual phenotype, given its mosaic distribution. One way of doing this is to co-inject RNA for β-gal so that its distribution can be visualized at a later stage. The experimental RNA and β-gal RNA co-segregate to a large extent, so that their eventual distributions will be very similar (Amaya *et al.*, 1993). The use of RNA injection and β-gal tracking can be illustrated by some recent experiments performed in our laboratories designed to examine the role of fibroblast growth factor in patterning the zebrafish embryo during blastula and gastrula stages (Fig. 1–Griffin *et al.*, 1995). In these experiments synthetic RNA encoding the secreted Xenopus eFGF protein following injection into the fertilized egg causes posteriorization of the embryo and the generation of multiple tail buds. This correlates with the widespread induction of a number of transcription factors known to affect embryonic patterning, including *no-tail*. *No-tail* is thought to be important in the induction and patterning of mesoderm

Fig. 1. The analysis of FGF function in the zebrafish embryo. A. A normal zebrafish embryo at 19 hours. B. A comparable staged embryo which had been injected with RNA encoding Xenopus eFGF at the fertilized egg. The embryo forms a series of tail bud extensions (arrows) only one of which has a notochord (n). C. Normal expression of the *ntl* gene at the onset of gastrulation as revealed by *in situ* hybridization. gr: the germ ring. D. Overexpression of FGF causes expression of *ntl* throughout the gastrula. E. Normal zebrafish larva at 2 days of age. 1 represents the otic vesicle, 2 the first somite and 3 the cloaca. F. A comparable staged larva developed from an egg which had been injected with the mutant dominant negative FGF receptor RNA. G. Normal gastrula showing the expression of *ntl*, viewed from the animal pole. H. An embryo injected at the single cell stage with the FGF dominant negative receptor at a comparable stage to that shown in G showing lack of expression of *ntl* in part of the germ ring (arrows).

and formation of the posterior axis. The dramatic alteration in convergence–extension movements during gastrulation is consistent with this posteriorization. By contrast, injecting the RNA encoding a dominant negative form of the FGF receptor (Amaya, Musci & Kirschner, 1991) causes loss of posterior parts of the embryo as the FGF signalling pathway is interrupted. A comparison of this phenotype with the *no-tail* mutant shows them to be similar, but the FGFR dominant negative phenotype is more severe. That FGF and *no-tail* are functionally linked is indicated by the observation in the embryos injected with dominant negative FGFR that *no-tail* expression is absent where the injected RNA is located at the onset of gastrulation. This example serves to illustrate how the function of a signalling pathway can be dissected using RNA injection.

The future

The near future holds great promise for work on the zebrafish because of the flood of interesting mutations emerging from the large-scale screens. The continued generation of a genetic map is a crucial development, and may allow the first positional cloning projects to be started. Other techniques are being tackled, such as the establishment of a mouse ES cell equivalent, which is yet to be obtained in any other vertebrate, and the use of viruses for transgenesis. The future is becoming increasingly more promising for the studies of developmentally regulated genes in this organism.

References

Abdelilah, S., Solnica-Krezel, L., Stainier, D. & Driever, W. (1994). Implications for dorsoventral axis determination from the zebrafish mutant janus. *Nature*, **370**, 468–71.

Alexandre, D., Clarke, J.D.W., Oxtoby, E., Yan, Y-L., Jowett, T. & Holder, N. (1996). Ectopic expression of Hoxa-1 in the zebrafish alters the fate or the mandibular arch neural crest and phenocpies a retanoic acid induced phenotype. *Development*, (In Press).

Amaya, E., Musci, T. & Kirshner, M. (1991). Expression of a dominant negative mutant of the FGF receptor disrupts mesoderm formation in *Xenopus* embryos. *Cell*, 66, 257–70.

Amaya, E., Stein, P., Musci, T. & Kirschner, M. (1993). FGF signalling in the early specification of mesoderm in *Xenopus*. *Development*, **118**, 477–87.

Culp, R., Nusslein-Volhard, C. & Hopkins, N. (1991). High frequency germ line transmission of plasmid DNA sequences injected into fertilised zebrafish eggs. *Proceedings of the National Academy of Sciences, USA*, **88**, 7953–7.

Driever, W., Stemple, D., Schier, A. & Solnica-Krezel, L. (1994). Zebrafish: genetic tools for studying vertebrate development. *Trends in Genetics*, **10**, 152–9.

Gossler, A., Joyner, A., Rossant, J. & Skarnes, W. (1989). Mouse embryonic stem cells and reporter constructs to detect developmentally regulated genes. *Science*, **244**, 463–6.

Griffin, K., Patient, R. & Holder, N. (1995). Analysis of FGF function in normal and NoTail zebrafish embryos reveals separate mechanisms for the formation of the trunk and the tail. *Development* (in press).

Hatta, K., Kimmel, C.B., Ho, R. & Walker, C. (1991). The cyclops mutation blocks specification of the floor plate of the zebrafish CNS. *Nature*, **350**, 339–41.

Helde, K., Wilson, E., Cretekos, C. & Grunwald, D. (1994). Contribution of early cells to the fate map of the zebrafish gastrula. *Science*, **265**, 517–20.

Kimmel, C.B. (1989). Genetics and early development of the zebrafish. *Trends in Genetics*, **5**, 283–8.

Kimmel, C.B. & Warga, R. (1986). Tissue specific cell lineages originate in the gastrula of the zebrafish. *Science*, **231**, 365–8.

Kimmel, C.B. & Warga, R. (1988). Cell lineage and developmental potential of cells in the zebrafish embryo. *Trends in Genetics*, **4**, 68–74.

Kimmel, C.B., Kane, D., Walker, C., Warga, R. & Rothman, M. (1989). A mutation that changes cell movement and cell fate in the zebrafish embryo. *Nature*, **337**, 358–62.

Kimmel, C.B., Warga, R. & Schilling, T. (1990). Origin and organisation of the zebrafish fate map. *Development*, **108**, 581–94.

Krauss, S., Maden, M., Holder, N. & Wilson, S. (1992). Zebrafish pax (b) is involved in the formation of the midbrain/hindbrain boundary. *Nature*, **360**, 87–9.

Kreig, P. & Melton, D. (1984). Functional messenger RNAs are produced by SP6 *in vitro* transcription of cloned cDNAs. *Nucleic Acid Research*, **12**, 7057–70.

Macdonald, R., Barth, A., Xu, Q., Holder, N., Mikkola, I. & Wilson, S. (1995). Midline signalling is required for *Pax* gene regulation and pattering of the eyes. *Development* (in press).

Mullins, M., Hammerschmidt, M., Haffter, P. & Nusslein-Volhard, C. (1994). Large scale mutagenesis in the zebrafish: in search of genes controlling development in a vertebrate. *Current Biology*, **4**, 189–202.

Neave, B., Rodaway, A., Wilson, S.W., Patient, R. & Holder, N. (1995). Expression of zebrafish GATA 3 (gta 3) during gastrulation and neurulation suggests a role in the specification of cell fate. *Mechanical Development*, **51**, 169–82.

Nieto, M., Gilardi-Hebenstreit, P., Charnay, P. & Wilkinson, D. (1992). A receptor protein tyrosine kinase implicated in the segmental patterning of the hindbrain and mesoderm. *Development*, **116**, 1137–50.

Postlethwait, J., Johnson, S., Midson, C., Talbot, W., Gates, M., Ballinger, E., Africa, D., Andrews, R., Carl, T., Eisen, J., Horne, S., Kimmel, C., Hutchinson, M., Johnson, M. & Rodriguez, A. (1994). A genetic linkage map for the zebrafish. *Science*, **264**, 699–702.

Reinhard, E., Nedivi, E., Wegner, J., Skene, J. & Westerfield, M. (1994). Neural selective activation and temporal regulation of a mammalian GAP-43 promoter in zebrafish. *Development*, **120**, 1767–75.

Stuart, G., Vielkind, J., McMurray, J. & Westerfield, M. (1990). Stable lines of transgenic zebrafish exhibit reproducible patterns of transgene expression. *Development*, **109**, 577–84.

Schulte-Merker, S., van Eeden, F., Halpern, M., Kimmel, C. & Nusslein-Volhard, C. (1994). *No tail (nt)* is the zebrafish homologue of the mouse T (*brachyury*) gene. *Development*, **120**, 1009–15.

Solnica-Krezel, L., Schier, A. & Driever, W. (1994). Efficient recovery of ENU-induced mutations from the zebrafish germline. *Genetics*, **136**, 1401–20.

Strehlow, D. & Gilbert, W. (1993). A fate map for the first cleavage stages of the zebrafish. *Nature*, **361**, 451–3.

Wilson, E., Helde, K. & Grunwald, D. (1993). Something's fishy here–rethinking cell movements and cell fate in the zebrafish embryo. *Trends in Genetics*, **10**, 348–52.

Wittbrodt, J. & Rosa, F. (1994). Disruption of mesoderm and axis formation in fish by ectopic expression of activin variants: the role of maternal activin. *Genes and Development*, **8**, 1448–62.

Xu, Q., Alldus, G., Holder, N. & Wilkinson, D. (1995). Expression of truncated Sek receptor tyrosine kinase disrupts the segmental restriction of gene expression in the *Xenopus* and zebrafish hindbrain. *Development* (in press).

S. ENNION

Myosin heavy chain isogene expression in carp

Introduction

The myosin heavy chain (MyoHC) plays a central role in cell motility and muscle contraction of all species, and a better understanding of the molecular diversity present between different isoforms of this protein is necessary if we are to understand the fine tuning of locomotion at the molecular level. In skeletal muscle the myosin molecule is located in the thick filament of the myofibril and together with actin takes part in the mechanism by which the chemical energy of ATP is converted into mechanical work. In its native state, striated muscle myosin is a hexameric protein consisting of two 'heavy' polypeptide chains of approximately 220 kD each and four 'light' chains approximately 17–20 kD each (Weeds & Lowey, 1971). Both the heavy and light chains of myosin exist as multiple isoforms which produce distinct isoforms of myosin all with similar, but not identical, structure and function (Gros & Buckingham, 1987).

The two myosin heavy chains intertwine at their more carboxyl regions forming an α-helical coiled-coil, about 150 nm long, termed the rod region of the molecule. The NH_2 region of each individual MyoHC molecule consists of a globular head about 19 nm long and 5 nm wide (at the widest point) such that the myosin molecule as a whole contains two head regions. The head region of the MyoHC forms the 'crossbridge' between the thick and thin filaments of the sarcomere. It contains an actin binding domain, the sites where the myosin light chains interact and the region where the ATPase activity of the molecule is localized (Craig, 1986).

Although extensive polymorphisms of the myosin light chains have been shown in a variety of species, the predominant contributor to the functional diversity of muscle fibres is thought to be the MyoHC. Like many other contractile proteins, the MyoHCs are encoded by a highly conserved multigene family (Nguyen *et al.*, 1982). There are thought to exist at least eight separate striated muscle *MyoHC* genes in mammals and as many as 31 in chicken (Robbins *et al.*, 1986). Gerlach *et al.* (1990) estimated the number of MyoHC genes present in the carp to be as many as 28. No pseudogenes have yet been identified in any species studied, although the total number of striated muscle *MyoHC* genes which have been cloned and characterized is fewer than nine for any one species.

The process of alternative exon splicing has not been observed in the *MyoHC* genes of vertebrate striated muscle, and it is thought that in such species each individual MyoHC isoform at the protein level is encoded by a separate gene (Wydro *et al.*, 1983; Leinwand *et al.*, 1983). In *Drosophila*, however, there are two *MyoHC* genes, one of which codes for a non-muscle isoform and one a muscle-specific isoform, which can be alternatively spliced to give multiple transcripts (Rozek & Davidson, 1983; Bernstein *et al.*, 1983; Kiehart *et al.*, 1989; Kronert *et al.*, 1991).

Due to their large size, relatively few vertebrate *MyoHC* genes have been sequenced completely. Those vertebrate *MyoHC* genes which have been entirely sequenced are very similar in size at both the genomic and RNA levels, where they are approximately 24 kb and 6000 nucleotides, respectively, and show a high degree of organizational and sequence homology. Both the human β cardiac (Jaenicke *et al.*, 1990) and the chicken embryonic (Molina *et al.*, 1987) *MyoHC* genes have a total of 40 exons, whilst the rat embryonic *MyoHC* gene (Strehler *et al.*, 1986) has 41 exons. However, the exon positions in all three genes differs slightly in a minority of cases. The equivalent of exon 37 in the human β cardiac gene is split into two exons in the chicken and rat embryonic genes such that the position of the additional intron is identical in both of the latter two genes. Also, the intron separating the final two exons of the human β cardiac (exons 39 and 40) and the rat embryonic (exons 40 and 41) is absent in the chicken embryonic gene.

Whilst much work has been carried out characterizing the MyoHC isoform family in mammals and chicken, comparatively little work has been done to characterize the MyoHC isoforms present in fish species. The functional demands placed on the locomotor musculature of fish differ in many ways from those of terrestrial animals, not least, the temperature range over which the muscle has to function. Therefore, as well as allowing interesting cross-species comparisons, further information regarding MyoHC polymorphism in fish would also provide a valuable insight into the molecular mechanisms employed by fish in muscle adaptation during development and in adult life.

The carp (*Cyprinus carpio*) was chosen as an experimental subject in our studies since this species, like many other cyprinids, has a remarkable ability to adapt the contractile characteristics of skeletal muscle in response to seasonal changes in environmental temperature. Earlier attempts to elucidate MyoHC polymorphisms in fish species at the protein electrophoretic level have been limited by the inherent instability of fish myosin (Connell, 1960) and similar overall charge and mass of the isoforms leading to co-migrations (Martinez *et al.*, 1989, 1990*a,b*; Karasinski, 1993). It was therefore decided to conduct our investigation of MyoHC polymorphism in carp at the level of *MyoHC* gene expression. The initial aim of this work was to obtain specific gene probes for individual isoforms of the MyoHC in carp. Once isolated, these probes could be used to investigate tissue-specific, developmental and adaptational expression patterns of MyoHC isoforms in carp.

Isolation of carp *MyoHC* gene sequences

Extensive sequence homology exists between *MyoHC* genes. This homology is present between species as diverse as nematode and rat, and is especially noticeable between different isoforms within the same species. The sequence of the globular head region (S1) of the molecule is more highly conserved between species than the rod region (Strehler *et al.*, 1986). Highly conserved sequences within the S1 region include the ATP binding site, the region of the two active thiols and sequences in the 50 kD domain thought to be involved in actin binding. Divergent regions also exist in the S1 head which include the hinge region and the 80 amino acid *N*-terminal sequence (Strehler *et al.*, 1986).

Comparisons of the light meromyosin nucleotide sequences of five members of the chicken fast MyoHC family showed that extremely high homology exists within this family (Moore *et al.*, 1992). All of the chicken sequences studied showed greater than 90% homology to each other, and two of the most similar isoforms shared complete nucleotide identity in two regions of over 250 base pairs. The high sequence homology between MyoHC isoforms means that nucleotide probes to be used for the characterization of the expression patterns of different isoforms within the same species have to be chosen with great care. Large probes which cover coding region of the gene have been shown to cross-hybridize to a number of MyoHC isoforms within the same species (Eller *et al.*, 1989*b*; Stedman, Kelly & Rubinstein, 1990) and are therefore of limited use in determining the expression patterns of an individual isoform. The 3′ untranslated region (3′UTR) of the gene, however, shows considerable divergence between isoforms, and this region has been extensively used in hybridization experiments to characterize the expression of individual isoforms (Morgan & Loughna, 1989; Sutherland *et al.*, 1991; De Nardi *et al.*, 1993). Our initial aim in the work on the carp *MyoHC* isogene family was therefore to isolate 3′UTRs from individual isoforms. Once sub-cloned, these regions could be used as isogene-specific probes to facilitate the characterization of the expression patterns of different MyoHC isoforms in carp.

Gerlach *et al.* (1990) isolated 28 MyoHC λ clones from a carp genomic library by hybridization with two rabbit cDNA clones; pMHCβ174 which is a slow MyoHC-specific probe and pMHC20–40 which is a fast MyoHC-specific probe. Turay (1991) sequenced various randomly chosen fragments derived from these carp genomic λ clones and generated partial sequence information for coding regions of carp MyoHC genes. One of these sequenced fragments, a 1.1 kb *Hind* III fragment from the clone λFG2, was shown to contain sequence from intron 39 and part of exon 40, the penultimate exon of the gene making it possible to locate and sequence the 3′UTR from this gene. In order to identify other clones containing the 3′UTR of *MyoHC* genes, the 1.1 kb *Hind*III fragment from the clone λFG2 was used to probe Southern blots containing restriction digest products of the remaining 27 uncharacterized genomic clones. In this way a further two distinct *MyoHC* gene 3′ UTR sequences were isolated from the genomic clones λFG19 and λFG17.

In addition to analysis of the carp genomic clones, the rapid amplification of cDNA ends polymerase chain reaction (RACE PCR) (Frohman, Dush & Martin, 1988) method was utilized to amplify 3′ sequences of different *MyoHC* isogenes from carp muscle. Comparisons of MyoHCC-terminal amino acid sequences show that the penultimate exon of the gene (the equivalent to exon 40 in the rat *MyoHC* gene (Strehler *et al.*, 1986)) is well conserved between different isoforms both with the same species and between different species. An oligonucleotide FG2EXN40 (5′ AGGAAGGTCGAGCATGAACTGGAGG 3′) was synthesized, based on the nucleotide sequence of part of exon 40 of the *MyoHC* gene contained in the λFG2 clone, and RACE PCR was performed on cDNA prepared from red and white muscle from carp of different ages and also at different acclimation temperatures. A total of 20 clones derived from RACE PCR were sequenced and a total of five distinct MyoHC sequence types could be distinguished. All the clones derived from adult carp muscle (white and red muscle from warm and cold acclimated carp) contained the same MyoHC isoform (arbitrarily named type 1). A clone from 14-month old carp (designated as type 2) was shown to contain a MyoHC sequence identical to the isoform contained in the genomic clone λFG17. Unhatched, fertilized carp eggs yielded two different MyoHC sequences designated as types 3 and 4. These two sequence types were also isolated from 60-day-old carp fry at both warm and cold temperatures. An additional MyoHC sequence type, designated as type 5, was also isolated from 60-day-old carp fry acclimated to 28 °C. None of the PCR isolated clones showed identical sequence homology to the genomic clones λFG2 and λFG19. These two clones were therefore classed as belonging to an additional two types of *MyoHC* gene (types 7 and 6, respectively). Thus, by combining the methods of genomic clone analysis and 3′ RACE PCR, a total of seven distinct MyoHC sequences from which isoform-specific 3′ UTR probes could be generated were isolated. It should be noted that the numerical nomenclature adopted here for the separate carp *MyoHC* gene types is purely arbitrary and does not relate in any way to the nomenclature of *MyoHC* genes adopted for mammals or the nomenclature adopted for the gross classification of myosin protein families (Goodson & Spudich, 1993).

Analysis of sequence data from the 3′ ends of carp *MyoHC* genes

Sequence identity values (Table 1) demonstrate the divergence of the 3′UTR sequence between the different types of carp MyoHC isoforms isolated. The sequence identities for the 3′UTR sequence between two isoforms ranges from 34% between type 4 and type 6 to 85% between type 2 and type 7. The sequence identities of the coding regions presented ranges from 83% between type 1 and types 4 and 5 to 99% between type 4 and type 5. Comparisons of the 3′UTR sequences from the carp *MyoHC* genes with those from other species (Fig. 1) demonstrate sequence homology relationships between the 3′UTR sequences of MyoHC isoforms across different species. The mammalian 3′UTR sequences show higher homology between equivalent isoforms across different species than between other isoforms from the same species. In contrast, the 3′UTR sequences from carp, *Xenopus* and chicken show more homology intraspecies than interspecies. Interestingly, two of the carp 3′UTR sequences show more homology to the *Xenopus* MyoHC 3′UTRs than the other carp isoforms. However, this homology is low at 40% sequence identity. The type 3 carp MyoHC is positioned alone in the dendrogram indicating that it does not have any homology with the other sequences. The homologies between the 3′UTR sequence are likely to reflect evolutionary relationships between species, and it will be interesting in the future to compare the carp, chicken and

Table 1. *Sequence identity between the 3′ ends of seven carp MyoHC isoforms*

Type 1	λFG17 Type 2	Type 3	Type 4	Type 5	λFG19 Type 6	λFG2 Type 7	
	74%	37%	39%	43%	58%	73%	Type 1
87%		43%	35%	40%	63%	85%	Type 2
86%	88%		42%	44%	40%	39%	Type 3
83%	89%	96%		65%	34%	41%	Type 4
83%	88%	95%	99%		39%	52%	Type 5
90%	93%	86%	86%	85%		60%	Type 6
92%	94%	89%	90%	89%	95%		Type 7

Nucleotide sequences were aligned using the CLUSTAL DNA alignment program (Higgins & Sharp, 1988) (parameters = open gap cost of 10, unit gap cost of 10) and the percentage identities between each pair of sequences shown in the Table. The top right-hand diagonal of the table presents the sequence identities between the 3′ untranslated region sequences (from the stop codon to the first 'A' of the poly (A) tail). For the genomic clones λFG2 and λFG19 the site of the poly (A) tail was estimated by comparison with the other five types of isoform. The bottom left-hand diagonal of the Table presents sequence identities of the last two coding exons of the gene (exons 40 and 41). The intronic sequences of the genomic clones λFG2 and λFG19 were omitted for the purpose of this alignment.

Fig. 1. Dendrogram alignment of MyoHC 3′UTR sequences from a variety of species. Sequence data from the first nucleotide of the stop codon to the first A of the Poly(A)$^+$ tail were aligned using the CLUSTAL alignment program of Higgins & Sharp (1988) with an 'open gap' cost of 10 and a 'unit gap' cost of 10. The EMBL accession numbers of the source sequences are as follows. 1) X03740 2) X72591 3) S. Ennion unpublished results 4) G. McCoy personal communication. 5) K00988 6) X72590 7) G. McCoy personal communication 8) J00892 9) M12086 10) M16557 11) Z32858 12) S. Ennion unpublished results 13) G. McCoy personal communication 14) X72589 15) M74753 16) X04267 17) X13988 18) Y008821 19) Z34887 20) M12289 21) X05004 22–25) this study 26) M19932 27) M30605 28) X07273 29) X15939 30) M28654 31) D00943 32) K01867 33) X15938 34) M27235 35) M27237 36) M27234 37) M27238 38) M27238 39–41) this study.

Xenopus MyoHC sequences with isoforms from other species of fish, birds and amphibians to see if a relationship similar to that present in the mammalian isoforms is also present in these species.

The 3′UTR sequences of the rat fast MyoHC isoforms (IIa, IIb and IIx) have been shown to contain small homologous stretches of sequence (De Nardi *et al.*, 1993) and these have also been highlighted

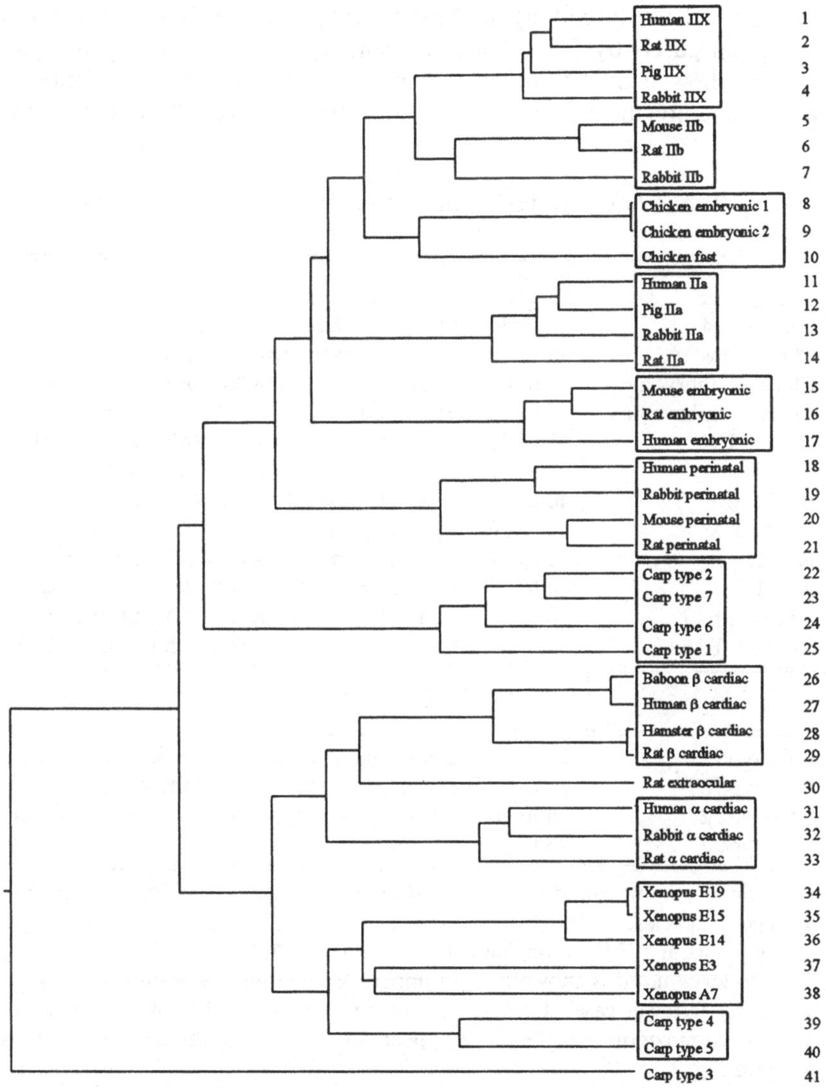

between the human and rat fast MyoHCs (Saez & Leinwand, 1986). The MyoHC sequences aligned to produce the dendrogram in Fig. 1 also showed some regions of homology between MyoHC isoforms. Whilst no sequence was found to be present in all the 3′UTR sequences aligned, a 10-nucleotide sequence 'AAAATGTGAA' was found to be present in 12 of the isoforms aligned (Table 2(*a*)). Moreover, this sequence was present in MyoHC 3′UTRs from a wide range of species where it appeared in approximately the same position (38–45 nucleotides after the stop codon) in the 3′UTR. It is unlikely that this motif is present purely by chance since the surrounding sequence in divergent species showed no homology. Furthermore, the fact that it occurs at approximately the same position after the stop codon in each gene

Table 2. *Conserved sequence motif in MyoHC 3′UTRs*

(*a*)

Isoform	Position	Sequence
Carp type 4	45 to 54	tttgtatctg **AAAATGTGAA** attgttcaat
Chicken fast	40 to 49	aggcatgcat **AAAATGTGAA** cctctgtgtt
Chicken embryonic 1	40 to 49	agaattgcac **AAAATGTGAA** attctatcac
Chicken embryonic 2	40 to 49	agaattgcac **AAAATGTGAA** attctatcac
Human IIX	41 to 50	agaaatgcac **AAAATGTGAA** aatctttgtc
Pig IIa	42 to 51	agagaggcac **AAAATGTGAA** gtctttgcgt
Rabbit IIb	40 to 49	agaaatgcac **AAAATGTGAA** gttcaaagtc
Rabbit IIX	40 to 49	agaaatgcac **AAAATGTGAA** actctttgtc
Rat IIa	42 to 51	agaaaggcac **AAAATGTGAA** gcctttggtc
Rat IIX	40 to 49	agagatgagc **AAAATGTGAA** gatctttgtc
Xenopus E19	38 to 47	tgaaatttgc **AAAATGTGAA** tttctcttcc
Xenopus E15	38 to 47	tgaaatttgc **AAAATGTGAA** tttcttccct

(*b*)

Carp type 1	21 to 35	aaactagaca **TACAAGCAAGCATAT** gactgacttg
Carp type 6	21 to 35	ataccacacc **TACAAGCAAGCATAT** aataagactg
Carp type 2	26 to 40	acaccacatc **TACAAGCAAGCATAT** aatatgactt
Carp type 7	26 to 40	agaccacatc **TACAAGCAAGCATAT** aatatgactt

(*a*) Sequence motif found in the 3′UTR across a range of MyoHC isoforms in a range of species.
(*b*) Sequence motif found in four of the seven carp MyoHC 3'UTRs.
The sequence motif is shown in bold upper case text and the flanking sequences in lower case. 'Position' corresponds to the number of nucleotides after the stop codon that the motif appears. The source of the sequence data is given in Fig. 1.

suggests that it has a functional role which has caused its conservation through evolution. Areas of sequence identity were also present in the carp MyoHC 3′UTR sequences, the most noticeable of which was a 15-nucleotide motif which was present in four of the seven carp isoforms (Table 2(*b*)).

The significance of these conserved sequence motifs in the 3′UTR is unclear at this time and requires further investigation. Band shift assays with cytoplasmic extracts from muscle cells would need to be performed in order to establish whether such motifs are binding recognition sequences for proteins which may be involved in RNA processing.

Determination of expression patterns of the carp MyoHC isoforms by Northern and *in situ* hybridization

With the sequence data obtained from the 3′ ends of the seven carp *MyoHC* genes isolated, it was possible to generate isoform specific probes for Northern hybridization which contained only sequence from the 3′UTR. In some cases the 3'UTR was sub-cloned adjacent to T7 and T3 RNA polymerase promoter sites in a plasmid vector to facilitate cRNA transcription of sense and antisense probes for *in situ* hybridizations.

Results from the Northern hybridization analysis performed with 3′UTR-containing probes are summarized in Fig. 2 with specific examples given in Figs. 4 and 5. The hybridization analysis performed demonstrates that the expression of carp *MyoHC* isogenes is both muscle fibre type and developmental stage specific. The expression of some of the isoforms also seems to be influenced by environmental temperature; however, in the case of their expression during development it is difficult to determine whether changes in their expression patterns are due to temperature *per se* or due to changes in the developmental stage induced by faster growth.

In mammals there is known to be a complex pattern *MyoHC* gene expression during muscle growth and development. The primary muscle fibres initially express the embryonic (Molina *et al.*, 1987; Eller *et al.*, 1989*b*; Karsch-Mizrachi *et al.*, 1989; Strehler *et al.*, 1986), neonatal (termed perinatal in human) (Periasamy, Wieczorek & Nadal-Ginard, 1984; Feghali & Leinwand, 1989; Weydert *et al.*, 1987) and the slow β cardiac (Barbet, Thornell & Butler-Brown, 1991; Narusawa *et al.*, 1987) MyoHC isoforms, but none of the fast MyoHC isoforms. The secondary fibre population of myotubes express embryonic, neonatal, and fast MyoHC isoforms in a very heterogeneous manner, but never the slow twitch MyoHC isoform (at least in humans) (Barbet *et al.*,

	Adult carp warm (28 °C) acclimated		Adult carp cold (10 °C) acclimated		Juvenile carp (12-14 months old)		Carp fry (60 day) (Whole fish)	
Isoform	White Muscle	Red Muscle	White Muscle	Red Muscle	White Muscle	Red Muscle	(28 °C)	(16 °C)
Type 1	*	-	*	-	*	-	*	-
Type 2	-	-	-	-	*	n	*	*
Type 3	-	-	-	-	*	-	*	*
Type 4	-	-	-	-	-	n	*	*
Type 5	*	*	*	*	*	n	*	*
Type 6	*	-	-	-	-	n	-	-
Type 7	*	-	-	-	-	n	-	-

Fig. 2. Summary of the expression patterns of seven carp MyoHC isoforms.
(*) Indicates that expression of the isoform was detected by Northern hybridization under high stringency conditions. (-) Indicates that expression of the isoform was not detected by Northern hybridization under high stringency conditions. (n) Indicates that Northern hybridization of the isoform with this particular muscle type have not been carried out.

1991). Later in embryonic development (at about 35 weeks gestation in the human) the expression of embryonic isoform decreases and shortly after birth the expression of the neonatal isoform also disappears. Concomitantly with the elimination of the embryonic and neonatal isoforms, the adult slow and fast MyoHC isoforms begin to be expressed predominantly as the muscle begins to take on its adult phenotype. Also at this stage, certain fibres cease to express the slow β cardiac isoform and begin to express the adult fast isoforms (Barbet *et al.*, 1991). Furthermore, immunological studies (Hughes *et al.*, 1993) suggest that there are at least three different slow MyoHC isoforms expressed during development in humans and rats, suggesting that the complexities of MyoHC switching during development are not yet fully understood. The hybridization data obtained in this study with the carp MyoHC probes demonstrate that the situation regarding developmental isoforms in carp is also complex. Three of the seven *MyoHC* isogenes isolated (types 2, 3, and 4) were shown to be expressed exclusively in immature carp. Two of these isoform types (types 3 and 4) were isolated by RACE PCR from cDNA prepared from unhatched carp eggs, so it could be expected that these two isoforms would be involved in development. The type 2 MyoHC isoform was isolated from the genomic clone λFG17 and also by RACE PCR on juvenile carp cDNA.

Both types 3 and 4 MyoHC isoforms are expressed in carp fry but only the type 3 isoform is expressed in older carp 12–14 months of age. Therefore, at some stage in the first year of development the type 4 *MyoHC* isogene ceases to be expressed. In this respect the type 4 isoform appears to be analogous to the mammalian neonatal isoform. However, such analogies may not be completely relevant in species as widely divergent as fish and mammals.

Northern hybridization with the type 2 MyoHC 3'UTR probe detected expression of this isoform in the white muscle of 12-month old carp and in carp fry. No expression was detected in the red muscle of 12-month old carp or in any sample from adult carp (Fig. 4). *In situ* hybridization performed with the same probe on whole sections from 12-month old carp (Fig. 3(*c*)) and on carp fry (data not shown) showed the pink muscle fibre layer of developing carp strongly expresses this isoform. Therefore, in addition to the 'small' and 'large' pink muscle fibre types described by Akster (1985), there would appear to be a further developmental classification of pink muscle fibre types in the carp.

The type 1 MyoHC isoform was shown to be expressed in both young and adult carp (Fig. 4). In the adult and juvenile carp, expression was only present in the white muscle fibres with no detectable expression, by Northern hybridization, in the red muscle fibre layer. With the carp fry it was not possible to dissect out the red and white muscle fibre layers in order to detect whether the expression of the type 1 isoform is restricted to the white fibres during the early developmental stages. In order to determine the fibre type specificity of the type 1 MyoHC isoform in carp fry, the 3'UTR of this isoform will need to be sub-cloned into a plasmid vector with RNA polymerase sites to facilitate *in situ* hybridization experiments.

Although the type 1 MyoHC isoform was not detected in the red muscle fibres of adult carp by Northern blot analysis, RACE PCR did detect the presence of type 1 MyoHC transcripts in adult red muscle. The presence of these type 1 MyoHC transcripts could be due to a very low 'background' expression of this gene in red muscle fibres, which was not strong enough to be detected by Northern hybridization

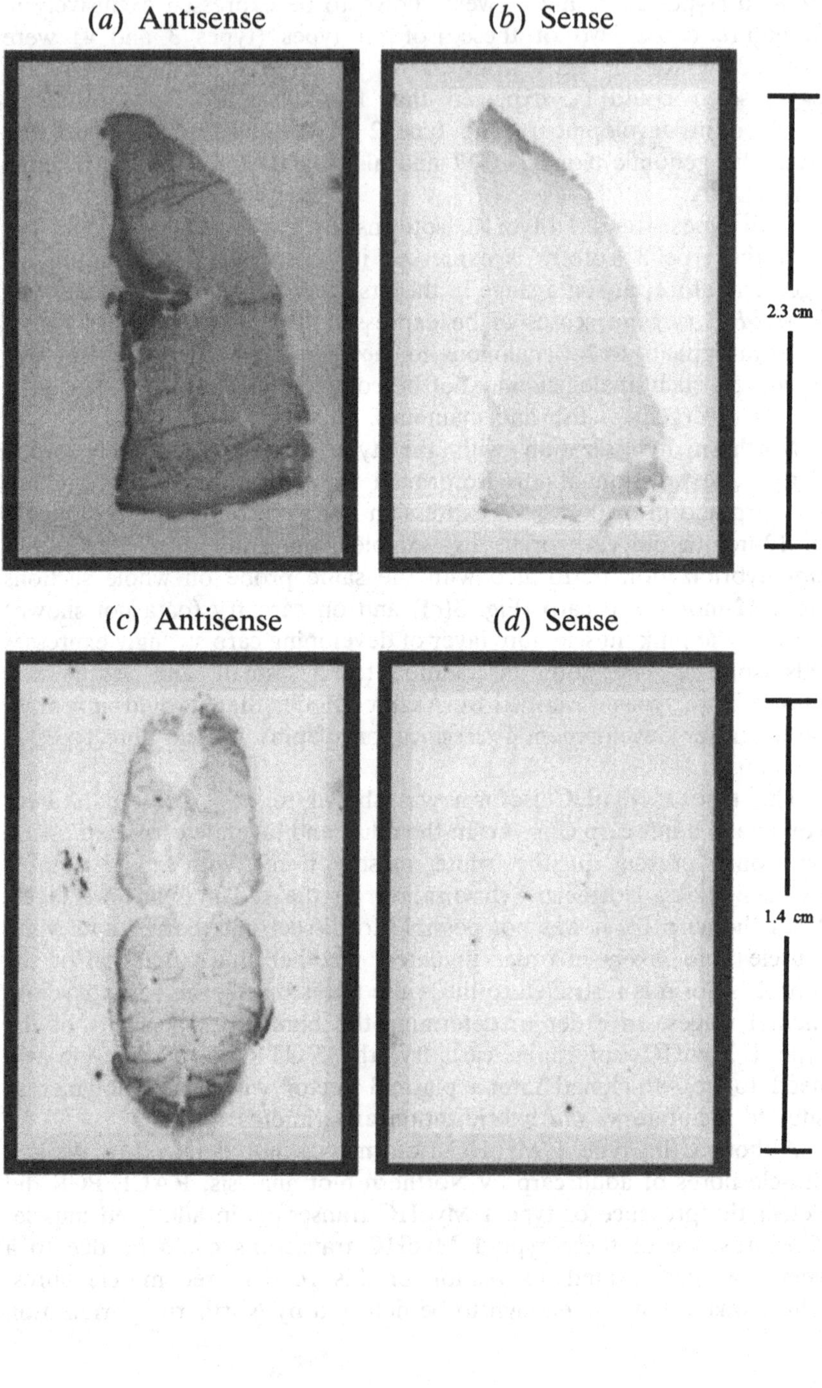
(a) Antisense
(b) Sense
2.3 cm
(c) Antisense
(d) Sense
1.4 cm

or alternatively the isolation of type 1 MyoHC from adult carp red muscle could be attributed to contamination of the red muscle sample with white muscle RNA during dissection. Due to the high sensitivity of the PCR, a contamination of only a few molecules of white muscle RNA would theoretically be enough to produce an amplification product.

Northern blot analysis of the expression of the type 1 isoform in carp fry could only detect expression in carp fry which had been acclimated to a warm temperature. Although the warm and cold acclimated fry used were from the same spawning, it is likely that the warm acclimated fish entered a more advanced state of development than the cold acclimated fry. Visual inspection and physical size of the carp fry after the temperature acclimation period supported this theory. Therefore, rather than being an acclimatory response to temperature, it is likely that this difference in expression of the type 1 MyoHC between the warm and cold acclimated fry is due to differences in developmental stage. From the results obtained in this study it can be concluded that at least five different isoforms of the MyoHC are expressed in carp skeletal muscle during the first year of development.

The type 5 MyoHC isoform was detected in all muscle types investigated from both young and adult carp. *In situ* hybridization experiments using the 3′ untranslated region probe of this isoform would be required to further investigate the expression of this isoform.

Fig. 3. *In situ* hybridization of ^{35}S labelled Type 2 MyoHC 3′UTR and HIIATP probes with carp muscle. (*a*) and (*b*) *In situ* hybridizations with ^{35}S labelled HIIATP cRNA sense and antisense probes on cryosections from muscle blocks taken from just below the dorsal fin, extending down the full flank, of four year-old adult carp (18–20 cm long). Note specific hybridization of the antisense probe to fibres within the red muscle block. (*c*) and (*d*) *In situ* hybridizations with ^{35}S labelled type 2 3′UTR MyoHC cRNA sense and antisense probes on cryosections (full cross-section of fish in ventral region) from 14 month-old juvenile carp. Note hybridization to the pink muscle fibre layer with the antisense probe. Prehybridization and hybridization conditions were the same as previously described (Ennion *et al.*, 1995). After posthybridization washes the sections were dehydrated and exposed to X-ray film for 1 week at 4 °C.

None of the seven isoform types isolated in this current study was shown to be expressed exclusively in the red muscle fibres of carp. Difference between myosin heavy chain proteins in red and white muscle have, however, been demonstrated at the protein electrophoretic level (Karasinski, 1993). The existence of at least one red muscle-specific MyoHC isoform in carp was also demonstrated at the mRNA level by the hybridization pattern of a probe (HIIATP) which contains the ATP binding site of the human β cardiac MyoHC isoform (a slow isoform). *In situ* hybridization at high stringency (Fig. 3(*a*)) shows that the HIIATP probe hybridizes specifically to the red muscle fibres. Therefore, the human β cardiac MyoHC isoform is more homologous to the MyoHC isoform(s) present in carp red muscle fibres than it is to the isoforms expressed in the white muscle fibres. Moreover, this

---

Fig. 4. Northern hybridization of carp RNA with 3′UTR probes for types 1 and 2 carp MyoHC. Total RNA (30 μg) extracted from various carp tissue samples was separated by electrophoresis and transferred to nylon membrane. The membrane was then hybridized, under high stringency conditions, with ^{32}P labelled Type 1 MyoHC 3′UTR probe. Panel (*a*) shows the autoradiograph (48 hours exposure) after this hybridization. The membrane was then stripped and subsequently rehybridized, under high stringency conditions, with ^{32}P labelled Type 2 MyoHC 3′UTR probe (panel (*b*), 72-hour exposure of autoradiograph). Panel (*d*) shows the original RNA gel stained with ethidium bromide. The RNA samples loaded in each lane are as follows. (1) White muscle from three juvenile carp (12 months of age, 5.5–6 cm in length). (2) Red muscle from the same fish in lane 1. (3) White muscle from three adult carp (4 years of age 19.0–22.0 cm in length), acclimated to 28 °C for 5 weeks. (4) White muscle from three adult carp (4 years of age 18.5–21.3 cm in length), acclimated to 16 °C for 5 weeks. (5) Adult carp spleen. (6) White muscle from three adult carp (4 years of age 16.2–19.1 cm in length), acclimated to 28 °C for 5 weeks. (7) Five whole carp fry (60 days old, 1.2–1.6 cm in length), acclimated to 20 °C for 5 weeks. (8). Red muscle from three adult carp acclimated to 28 °C for 5 weeks (same fish as in lane 3). (9) White muscle from three adult carp (4 years of age 18.3–21.2 cm in length), acclimated to 20 °C for 5 weeks. (10) Five whole carp fry (60 days old, 0.8–1.1 cm in length), acclimated to 16 °C for 5 weeks. (11) Red muscle from three adult carp acclimated to 28 °C for 5 weeks (same fish as in lane 6). (12) Red muscle from three adult carp acclimated to 16 °C for 5 weeks (same fish as in lane 4).

apparently high conservation of DNA sequence coding for the ATP binding site of the slow MyoHC isoforms in species as divergent as carp and human suggests that this region is very important in determining the functional properties of slow MyoHC isoforms.

The failure of the RACE PCR technique to yield a red muscle specific isoform is probably due to the choice of the 5′ oligonucleotide used in the PCR. This oligonucleotide (FG2EXN40) was designed from the sequence obtained from the genomic clone λFG2 and, since the isoform encoded in this particular clone proved to be a white muscle-specific MyoHC isoform, it is possible that the sequence of FG2EXN

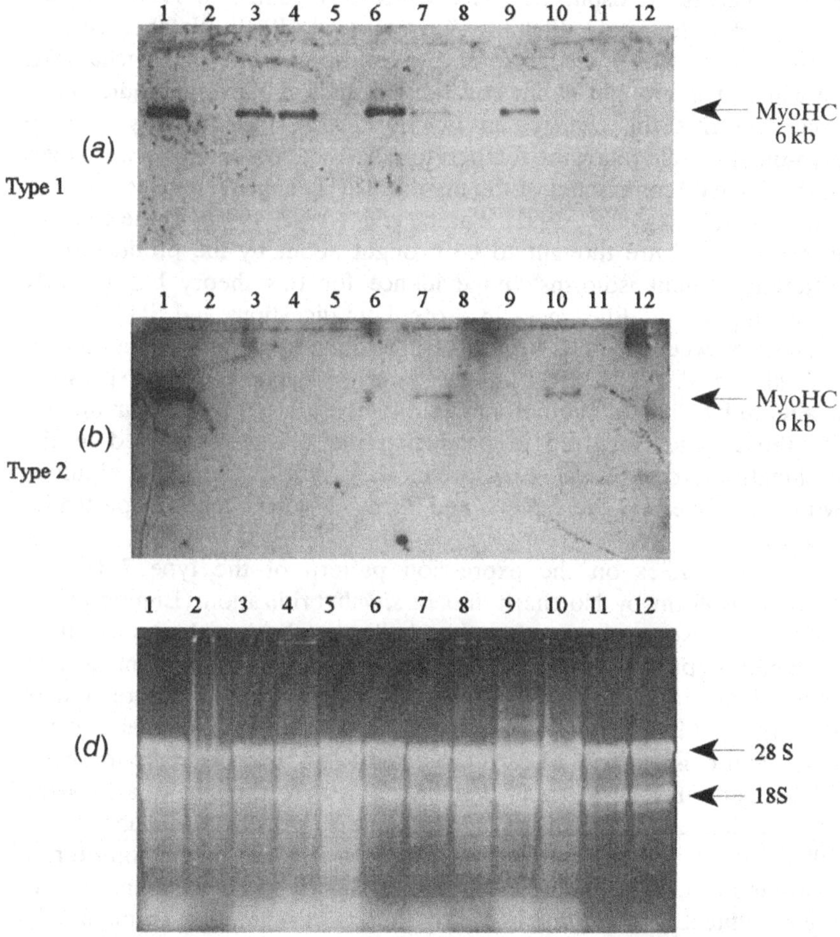

40 is not sufficiently homologous to the slow red MyoHC isoform to allow amplification.

Two MyoHC isoforms (Types 6 and 7) were shown to be expressed exclusively in the white muscle fibres of adult carp. Moreover, both isoform types were only detected in the white muscle from carp which had been acclimated to warm temperatures, indicating that these two isoform types may be involved in some way in the temperature acclimation process in carp.

Thermal acclimation allows carp and some other cyprinid fish to compensate for changes in environmental temperature which would otherwise be detrimental to muscle function. This thermal acclimation process has been well documented (Johnston, Sidell & Driedzic, 1985; Heap, Watt & Goldspink, 1987; Crockford & Johnston, 1990; Gerlach *et al.*, 1990; Johnston, Fleming & Crockford, 1990; Watabe, Guo & Hwang, 1994) (also see Chapter by Goldspink) but the mechanisms by which it is brought about still remain unclear. Several studies have suggested that the changes in myofibrillar ATPase activity in carp myotomal muscle following temperature acclimation result from changes in the MyoHC component of the myofibrils (Hwang, Watabe & Hashimoto, 1990; Hwang *et al.*, 1991; Watabe *et al.*, 1992, 1995). These changes in the MyoHC are thought to be brought about by the production of different protein isoforms and evidence for this theory has recently been supplied by differences in proteolytic digestions and SDS–PAGE analysis between MyoHC proteins from cold and warm acclimated carp (Watabe *et al.*, 1992, 1994). Hence, *MyoHC* genes whose expression is altered by changes in environmental temperature are likely candidates for those genes involved in producing the changes observed in the acclimation process making the preliminary hybridization data obtained with the probes for the types 6 and 7 carp MyoHC genes of particular interest.

Further studies on the expression pattern of the type 7 (λFG2) MyoHC isoform by Northern and *in situ* hybridization (Ennion *et al.*, 1995) demonstrate that, rather than being involved in the temperature acclimation process *per se*, this isoform is involved in temperature induced myotomal muscle growth by fibre hyperplasia. Expression of the type 7 MyoHC isoform is localized to the small diameter white myotomal muscle fibres of carp subjected to an increase in environmental temperature. These small diameter fibres are thought to arise from myosatellite cell recruitment during muscle growth (Koumans *et al.*, 1993*a,b*). Hence the pattern of MyoHC isoform expression during myosatellite cell recruitment in carp is slightly different from that in myosatellite cell recruitment during muscle regeneration in mammals, since in mammals the embryonic and neonatal MyoHC isoforms are re-expressed (Sartore, Gorzá & Schiaffino, 1982).

The type 6 carp *MyoHC* gene probe also shows a hybridization pattern in which expression is only observed in the white muscle of adult carp acclimated to a warm temperature (Fig. 5). No expression of this isoform was detected in carp acclimated to 10 °C or in carp raised to 20 °C for one week. However, after two weeks at 28 °C expression of this isoform could be detected in the white muscle but not the red. Hence, the carp type 6 *MyoHC* gene may be involved in the temperature acclimation response of carp white muscle to warm temperatures. As already highlighted in the case of the type 7 isoform, knowledge of which fibre population expresses the type 6 MyoHC isoform during acclimation to warm environmental temperature is essential to establish whether this isoform is involved in the temperature acclimation process or growth process, and *in situ* hybridization experiments using the 3′ untranslated region probe for this isoform are required.

Discussion

Whilst it is unlikely that the seven MyoHC sequences isolated in this study account for the full family of MyoHC isoforms in the carp, the molecular analysis performed gave an insight into the complexities of MyoHC polymorphism in a species of fish. The question of whether the seven MyoHC 3′UTR sequences isolated in this study correspond to seven distinct *MyoHC* genes, rather than some types being alleles of the same gene, was addressed by assessing both sequence comparison data and the expression patterns obtained by hybridization of 3′UTR-containing probes. Comparison of the deduced amino acid sequences corresponding to the last two exons of the MyoHC gene (Fig. 6), shows a high degree of homology between all seven of the MyoHC types isolated. Indeed, two of the carp MyoHC types (types 4 and 5) have identical deduced amino acid sequence over this region. Such high sequence homology in this region of the gene is not uncommon between different MyoHC isoforms within the same species. A similarly high sequence homology is also present in the rat MyoHC isoforms where the amino acid sequence of the fast IIa and IIb isoforms are identical to one another over the known carboxyl terminal region and the IIX isoform has only one amino acid difference (De Nardi *et al.*, 1993). Thus, for five of the seven MyoHC types deduced amino acid sequence differences can be taken as indicative that they correspond to separate genes, especially in view of the fact that no vertebrate striated muscle MyoHC gene to date has been found to undergo alternative splicing of exons.

The 3′UTR nucleotide sequences of the seven carp MyoHC types show, in most cases, sufficient sequence divergence from each other (Table 1) to strongly suggest that they correspond to separate genes. The two carp MyoHC types which showed the highest sequence homology in their 3′UTR sequence (85% identity) were type 7 and type 2. The homology between the 3′UTR sequences of these two MyoHC types is sifficiently high to raise suspicion that they correspond to alleles of the same *MyoHC* gene. However, two lines of evidence suggest that these two MyoHC types do, in fact, correspond to distinct genes. Firstly, analysis of the corresponding genomic clones revealed that the intron between the penultimate and final exon differs in both sequences

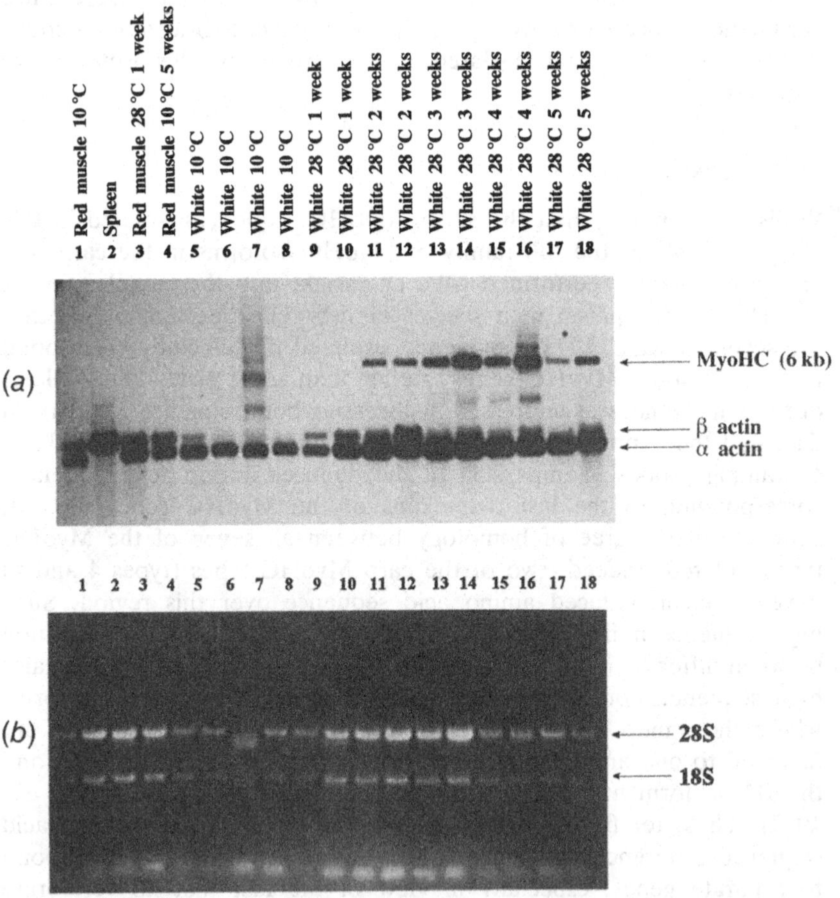

and length. Secondly, these two carp MyoHC types show a completely different pattern of expression when the 3′UTR is used as probe. The type 2 MyoHC 3′UTR probe hybridized exclusively to the pink muscle fibres in immature carp whereas the type 7 MyoHC 3′UTR probe hybridized only to the small diameter muscle fibres in adult carp acclimated to a warm temperature. Therefore, considering the results of sequence comparison analysis and expression studies, it is likely that all seven carp MyoHC types isolated in this study correspond to distinct MyoHC genes rather than some being alleles of the same gene.

The high sequence homology between the 3′UTR sequences of the type 7 and type 2 MyoHC isoforms suggests that evolutionary recent gene conversion events, similar to those which have been proposed for the chicken (Moore *et al.*, 1992, 1993) have also occurred in the carp. Furthermore, given the fact that these two isoform types have very different expression patterns, one might hypothesize that such a gene conversion event led to the splicing of the 3′ end of a common ancestral gene to different 5′ ends which have different regulatory (promoter) regions, hence the different expression patterns of the resulting dupli-

---

Fig. 5. Hybridization of the Type 6 MyoHC 3′UTR probe with RNA from carp maintained at warm and cold temperatures. Total RNA (30 μg) from carp acclimated to either 10 °C or 28 °C was separated by electrophoresis, transferred to nylon membrane and hybridized under high stringency conditions with the Type 6 MyoHC probe (Panel (*a*)). The blot was then exposed to X-ray film (Fuji RX) at −70 °C for 1 week. Subsequently the membrane was stripped and reprobed under high stringency conditions with the carp actin probe FGA101 (Gerlach *et al.*, 1990). The blot was exposed to X-ray film (Fuji RX) at −70 °C for 72 hours and superimposed on the autoradiograph of the FG19UTR hybridization. Panel (*b*) shows ethidium bromide staining of the original RNA agarose gel. Lanes are as follows. (1) Red muscle from four carp acclimated to 10 °C for five weeks. (2) Carp spleen. (3) Red muscle from four carp acclimated to 28 °C for one week. (4) Red muscle from four carp acclimated to 10 °C for five weeks. (5), (6), (7) and (8) White muscle from individual carp acclimated to 10 °C for five weeks. (9) and (10) White muscle from two carp acclimated to 28 °C for one week. (11) and (12) White muscle from two carp acclimated to 28 °C for two weeks. (13) and (14) White muscle from two carp acclimated to 28 °C for three weeks. (15) and (16) White muscle from two carp acclimated to 28 °C for four weeks. (17) and (18) White muscle from two carp acclimated to 28 °C for five weeks.

```
TYPE 1   RKVQHELEESHERADIAESQVNKLRAKSREAGKTKV EE
TYPE 2   RKVQHELEEAQERADVAESQVNKLRAKSRDAGKSKD EE
TYPE 3   RKVQHELEEAEERADIAESQVNKLRAKSRDAGKAK  EE
TYPE 4   RKVQHELEEAEERADIAESQVNKLRAKSRDAGKGKEAAE
TYPE 5   RKVQHELEEAEERADIAESQVNKLRAKSRDAGKGKEAAE
TYPE 6   RKVEHELEEAQERADIAESQVNKLRAKSRDAGKSKD EE
TYPE 7   RKVQHELEEAQERADIAESQVNKLRAKSRDAGKSKD EE
         ***.******.****.**************.***.*...*
```

Fig. 6. Deduced carboxy terminus amino acid sequence seven carp *MyoHC* genes. Amino acids identical in all seven isoforms are marked (*), position where differences occur are marked with (.).

cated genes. If such gene conversion events were responsible for the high sequence homology observed between the 3′UTR sequences of the carp type 2 and type 7 *MyoHC* genes, one would also expect to see a similarly high degree of homology in the intron sequence in the 3' ends of these two genes. Whilst only limited sequence data for these two *MyoHC* genes are available at this time, the intron between putative exons 40 and 41 has been sequenced for both these genes and this shows only limited sequence homology at the 3′ splice site. One possible explanation of why the 3′UTR sequence of the carp types 6 and 7 MyoHC isoforms should show more homology than the preceding introns is that the 3′UTR plays some role in the regulation of gene expression or RNA processing (Jackson & Standart, 1990; Rastinejad & Blau, 1993) and this role has led to selective pressure for the 3′UTR sequence to be more conserved than the intron sequence. More extensive sequence data from the genomic clones λFG2 and λFG17 would be required to further investigate the evolutionary relationship between these two isoforms. The promoter sequence of the λFG2 clone (type 7) has already been isolated, sequenced and partially characterized (see chapters by Müller and by Goldspink in this volume), and comparison of this sequence with the promoter region of λFG17 may yield some interesting data.

Future studies

The isolation of isoform-specific carp *MyoHC* gene probes in this study has provided the molecular tool for many possible future lines of investigation. Of particular interest would be to isolate the sequence coding of the ATPase and actin binding domains of the different isoforms. In this way it may be possible to begin to understand why so many isoforms of the MyoHC are necessary and particularly how different isoforms are more suited to a particular function in locomotion than others.

Acknowledgements

This work was funded by grants from the NERC and SERC.

References

Akster, H.A. (1985). Morphometry of muscle fibre types in the carp (*Cyprinus carpio* L.). *Cell and Tissue Research*, **241**, 193–201.

Barbet, J.P., Thornell, L.E. & Butler-Browne, G.S. (1991). Immunocytochemical characterisation of two generations of fibres during the development of the human quadriceps muscle. *Mechanisms of Development*, **35**, 3–11.

Bernstein, S.I., Mogami, K., Donady, J.J. & Emerson, C.P.Jr. (1983). Drosophila muscle myosin heavy chain encoded by a single gene in a cluster of muscle mutations. *Nature*, **302**, 393–7.

Craig, R. (1986). The structure of the contractile filaments. In *Myology, Basic and Clinical*, ed. A.G. Engel & B.Q. Banker, pp. 72–123. New York: McGraw-Hill.

Connell, J.J. (1960). Studies on the proteins of fish skeletal muscle. 7. Denaturation and aggregation of cod myosin. *Biochemical Journal*, **75**, 530–8.

Crockford, T. & Johnston, I.A. (1990). Temperature acclimation and the expression of contractile protein isoforms in the skeletal muscles of the common carp (*Cyprinus carpio*). *Journal of Comparative Physiology–Biochemical, Systemic and Environmental Physiology*, **160**, 23–30.

De Nardi, C., Ausoni, S., Moretti, P., Gorza, I., Velleca, M., Buckingham, M. & Schiaffino, S. (1993). Type-2x-myosin heavy chain is coded by a muscle fibre type-specific and developmentally regulated gene. *Journal of Cell Biology*, **123**, 823–35.

Eller, M., Stedman, H.H., Sylvester, J.E., Fertels, S.H., Rubinstein, N.A., Kelly, A.M. & Sarkar, S. (1989*a*). Nucleotide sequence of full length human embryonic myosin heavy chain cDNA. *Nucleic Acids Research*, **17**, 3591–2.

Eller, M., Stedman, H.H., Sylvester, J.E., Fertels, S.H., Wu, Q.L., Raychowdbury, M.K., Rubinstein, N.A., Kelly, A.M. & Sarkar, S. (1989*b*). Human embryonic myosin heavy chain cDNA. Interspecies sequence conservation of the myosin rod, chromosomal locus and isoform specific transcription of the gene. *FEBS Letters*, **256**, 21–8.

Ennion, S., Sant'ana Pereira, J., Sargeant, A.J., Young, A. & Goldspink, G. (1995). Characterization of human skeletal muscle fibres according to the myosin heavy chains they express. *Journal of Muscle Research and Cell Motility*, **16**, 35–43.

Feghali, R. & Leinwand, L.A. (1989). Molecular genetic characterization of a developmentally regulated human perinatal myosin heavy chain. *Journal of Cell Biology*, **108**, 1791–7.

Frohman, M.A., Dush, M.K. & Martin, G.R. (1988). Rapid production of full length cDNAs from rare transcripts: amplification using a single gene specific oligonucleotide primer. *Proceedings of the National Academy of Sciences, USA*, **85**, 8998–9002.

Gerlach, G.F., Turay, L., Malik, K.T., Lida, J., Scutt, A. & Goldspink, G. (1990). Mechanisms of temperature acclimation in the carp: a molecular biology approach. *American Journal of Physiology*, **259**, R237–44.

Goodson, H.V. & Spudich, J.A. (1993). Molecular evolution of the myosin family: relationships derived from comparisons of amino acid sequences. *Proceedings of the National Academy of Sciences, USA*, **90**, 659–63.

Gros, F. & Buckingham, M.E. (1987). Polymorphism of contractile proteins. *Biopolymers*, **26**, S177–92.

Heap, S.P., Watt, P.W. & Goldspink, G. (1987). Contractile properties of goldfish fin muscles following temperature acclimation. *Journal of Comparative Physiology–B, Biochemical, Systemic and Environmental Physiology*, **157**, 219–25.

Higgins, D.G. & Sharp, P.M. (1988). Clustal: a package for performing multiple sequence alignment on a microcomputer. *Gene*, **73**, 237–44.

Hughes, S.M., Cho, M., Karsch-Mizrachi, I., Travis, M., Silberstein, L. & Blau, H.M. (1993). Three slow myosin heavy chains sequentially expressed in developing mammalian skeletal muscle. *Developmental Biology*, **158**, 183–99.

Hwang, G.C., Watabe, S. & Hashimoto, K. (1990). Changes in carp myosin ATPase induced by temperature acclimation. *Journal of Comparative Physiology–B, Biochemical, Systemic and Environmental Physiology*, **160**, 233–9.

Hwang, G.C., Ochiai, Y., Watabe, S. & Hashimoto, K. (1991). Changes of carp myosin subfragment-1 induced by temperature acclimation. *Journal of Comparative Physiology–B, Biochemical, Systemic and Environmental Physiology*, **161**, 141–6.

Jackson, R.J. & Standart, N. (1990). Do the Poly(A) tail and 3' untranslated region control mRNA translation? *Cell*, **62**, 15–24.

Jaenicke, T., Diederich, K.W., Haas, W., Schleich, J., Lichter, P., Pfordt, M. & Vosberg, H.P. (1990). The complete sequence of the human beta-myosin heavy chain gene and a comparative analysis of its product. *Genomics*, **8**, 194–206.

Johnston, I.A., Fleming, J.D. & Crockford, T. (1990). Thermal acclimation and muscle contractile properties in cyprinid fish. *American Journal of Physiology*, **259**, R231–6.

Johnston, I.A., Sidell, B.D. & Driedzic, W.R. (1985). Force–velocity characteristics and metabolism of carp muscle fibres following temperature acclimation. *Journal of Experimental Biology*, **119**, 239–49.

Karasinski, J. (1993). Diversity of native myosin and myosin heavy chain in fish skeletal muscles. *Comparative Biochemistry and Physiology [B]*, **106**, 1041–7.

Karsch-Mizrachi, I., Travis, M., Blau, H. & Leinwand, L.A. (1989). Expression and DNA sequence analysis of a human embryonic skeletal muscle myosin heavy chain gene. *Nucleic Acids Research*, **17**, 6167–79.

Kiehart, D.P., Lutz, M.S., Chan, D., Ketchum, A.S., Laymon, R.A. & Nguyen, B. (1989). Identification of the gene for fly non-muscle myosin heavy chain: *Drosophila* myosin heavy chains are encoded by a gene family [published erratum appears in *EMBO J.* 1989 Jun;8(6):1896]. *EMBO Journal*, **8**, 913–22.

Koumans, J.T.M., Akster, H.A., Booms, G.H.R. & Osse, J.W.M. (1993*a*). Growth of carp (*Cyprinus carpio*) white axial muscle: hyperplasia and hypertrophy in relation to the myonucleus/sarcoplasm ratio and the occurrence of different subclasses of myogenic cells. *Journal of Fish Biology*, **43**, 69–80.

Koumans, J.T.M., Akster, H.A., Booms, R.G.H. & Osse, J.W.M. (1993*b*). Influence of fish size on proliferation of cultured myosatellite cells of white axial muscle of carp (*Cyprinus carpio* L.). *Differentiation*, **53**, 1–6.

Kronert, W.A., Edwards, K.A., Roche, E.S., Wells, L. & Bernstein, S.I. (1991). Muscle-specific accumulation of *Drosophila* myosin heavy chains: a splicing mutation in an alternative exon results in an isoform substitution. *EMBO Journal*, **10**, 2479–88.

Leinwand, L.A., Fournier, R.E., Nadal-Ginard, B. & Shows, T.B. (1983). Multigene family for sarcomeric myosin heavy chain in mouse and human DNA: localization on a single chromosome. *Science*, **221**, 766–9.

Manstein, D.J., Titus, M.A., De Lozanne, A. & Spudich, J.A. (1989). Gene replacement in Dictyostelium: generation of myosin null mutants. *EMBO Journal*, **8**, 923–32.

Martinez, I., Olsen, R.L., Ofstad, R., Janmot, C. & d'Albis, A. (1989). Myosin isoforms in mackerel (*Scomber scombrus*) red and white muscles. *FEBS Letters*, **252**, 69–72.

Martinez, I., Ofstad, R. & Olsen, R.L. (1990*a*). Electrophoretic study of myosin isoforms in white muscles of some teleost fishes. *Comparative Biochemistry and Physiology–B: Comparative Biochemistry*, **96B**, 221–7.

Martinez, I., Ofstad, R. & Olsen, R.L. (1990*b*). Intraspecific myosin light chain polymorphism in the white muscle of herring (*Clupea harengus harengus*, L.). *FEBS Letters*, **265**, 23–6.

Molina, M.I., Kropp, K.E., Gulick, J. & Robbins, J. (1987). The sequence of an embryonic myosin heavy chain gene and isolation of its corresponding cDNA. *Journal of Biological Chemistry*, **262**, 6478–88.

Moore, L.A., Tidyman, W.E., Arrizubieta, M.J. & Bandman, E. (1992). Gene conversions within the skeletal myosin multigene family. *Journal of Molecular Biology*, **223**, 383–7.

Moore, L.A., Tidyman, W.E., Arrizubieta, M.J. & Bandman, E. (1993). The evolutionary relationship of avian and mammalian myosin heavy-chain genes. *Journal of Molecular Evolution*, **36**, 21–30.

Narusawa, M., Fitzsimons, R.B., Izumo, S., Nadel-Ginard, B. & Rubinstein, N.A. (1987). Slow myosin in developing rat skeletal muscle. *Journal of Cell Biology*, **104**, 447–59.

Nguyen, H.T., Gubits, R.M., Wydro, R.M. & Nadal-Ginard, B. (1982). Sarcomeric myosin heavy chain is coded by a highly conserved multigene family. *Proceedings of the National Academy of Sciences, USA*, **79**, 5230–4.

Periasamy, M., Wieczorek, D.F. & Nadal-Ginard, B. (1984). Characterization of a developmentally regulated perinatal myosin heavy-chain gene expressed in skeletal muscle. *Journal of Biological Chemistry*, **259**, 13573–8.

Rastinejad, F. & Blau, H.M. (1993). Genetic complementation reveals a novel regulatory role for 3′ untranslated regions in growth and differentiation. *Cell*, **72**, 903–17.

Robbins, J., Horan, T., Gulick, J. & Kropp, K. (1986). The chicken myosin heavy chain family. *Journal of Biological Chemistry*, **261**, 6606–12.

Rozek, C.E. & Davidson, N. (1983). Drosophila has one myosin heavy-chain gene with three developmentally regulated transcripts. *Cell*, **32**, 23–34.

Saez, L. & Leinwand, L.A. (1986). Characterization of diverse forms of myosin heavy chain expressed in adult human skeletal muscle. *Nucleic Acids Research*, **14**, 2951–69.

Sartore, S., Gorzá, L. & Schiaffino, S. (1982). Fetal myosin heavy chains in regenerating muscle. *Nature*, **298**, 294–6.

Strehler, E.E., Strehler-Page, M.A., Perriard, J.C., Periasamy, M. & Nadal-Ginard, B. (1986). Complete nucleotide and encoded amino acid sequence of a mammalian myosin heavy chain gene. Evidence against intron-dependent evolution of the rod. *Journal of Molecular Biology*, **190**, 291–317.

Sutherland, C.J., Elsom, V.L., Gordon, M.L., Dunwoodie, S.L. & Hardeman, E.C. (1991). Coordination of skeletal muscle gene expression occurs late in mammalian development. *Developmental Biology*, **146**, 167–78.

Turay, L.R. (1991). Molecular aspects of temperature acclimation in the muscle of the carp (*Cyprinus carpio* L.), University of London: PhD thesis.

Watabe, S., Hwang, G.C., Nakaya, M., Guo, X.F. & Okamoto, Y. (1992). Fast skeletal myosin isoforms in thermally acclimated carp. *Journal of Biochemistry*, **111**, 113–22.

Watabe, S., Guo, X.F. & Hwang, G.C. (1994). Carp express specific isoforms of the myosin cross-bridge head, subfragment-1, in association with cold and warm temperature acclimation. *Journal of Thermal Biology*, **19**, 261–8.

Watabe, S., Imai, J.I., Nakaya, M., Hirayama, Y., Okamoto, Y., Masaki, H., Uozumi, T., Hirono, I. & Aoki, T. (1995). Temperature acclimation induces light meromyosin isoforms with different primary structures in carp fast skeletal muscle. *Biochemical and Biophysical Research Communications*, **208**, 118–25.

Weeds, A.G. & Lowey, S. (1971). Substructure of the myosin molecule II. The light chains of myosin. *Journal of Molecular Biology*, **61**, 701–25.

Weydert, A., Barton, P., Harris, A.J., Pinset, C. & Buckingham, M. (1987). Developmental pattern of mouse skeletal myosin heavy chain gene transcripts *in vivo* and *in vitro*. *Cell*, **49**, 121–9.

Wydro, R.M., Nguyen, H.T., Gubits, R.M. & Nadal-Ginard, B. (1983). Characterization of sarcomeric myosin heavy chain genes. *Journal of Biological Chemistry*, **258**, 670–8.

L. GAUVRY, C. PEREZ and
B. FAUCONNEAU

Rainbow trout myosin heavy chain polymorphism during development

Introduction

Trunk muscle in fish, is composed of three distinct muscles (white, red and pink) which are involved in the different swimming modes (cruise or escape). Myotomal muscle has been characterized by histo- or immunohisto-chemical methods in diverse fish species, but few investigations at the molecular level have been performed.

In fish, myosin, the major contractile muscular protein, presents the same structure as that found in higher vertebrates. It is a hexameric molecule constituted by the association of one pair of alkali light chains (LC2) and one pair of regulatory light chains (LC1 and/or LC3) with two myosin heavy chains (MyoHC). The molecule presents a bipolar structure: the head and the rod. The head results from a globular arrangement of the *N*-terminal region of the MyoHC or from the proteolytic subfragment-1 associated to the light chains. The head of the molecule contains the ATP binding site and the actin site necessary to the contraction caused by a sliding of the head with regard to the thick filament formed by the rods of myosin heavy chains. The rod corresponds to a coiled-coil arrangement of two α helical myosin heavy chains.

Differences in fibre types are due to different myosins with the expression of either slow or fast myosin heavy and light chains. Skeletal muscle in mammals has been shown to be composed of two slow MyoHCs (MyoHCI, MyoHCIton), four fast MyoHCs (MyoHCIIa, MyoHCIIb, MyoHCIId, MyoHCαcard) and two super fast MyoHCs (MyoHCIIm, MyoHCIIeom) (for review see Staron & Johnson, 1993). In addition, there are developmental MyoHC isoforms (embryonic, neonatal) related to muscle maturation and which appear sequentially during development (Whalen *et al.*, 1981). Persistence of developmental isoforms has been observed in the human masseter, in the rat and bovine extra-ocular or in the rat intra-fusal fibres (Soussi-Yanicostas *et al.*, 1990; Wieczorek *et al.*, 1985; Sartore *et al.*, 1987; Pedrosa *et al.*, 1989). The co-expression of different MyoHCs has been shown within the same muscle fibre, during a fibre-type transition induced by low frequency stimulations (Termin, Staron & Pette, 1989). In fish, a myosin polymorphism specific to fibre types has been demonstrated at the protein level and data are presented in this chapter.

At the molecular level, gene studies have shown that MyoHC is a multigene family and each MyoHC is encoded by a different gene. MyoHC isoforms present two hypervariable regions within the exons 7 and 17 of the subfragment-1. The last 10–15 amino acids as well as the untranslated region are divergent and peculiar for each MyoHC isoform (Radice & Malacinski, 1989). This multigene family also includes cardiac and smooth muscle MyoHCs and non-muscular MyoHCs (Buckingham, 1985). In fish, studies have been limited to carp (*Cyprinus carpio*), where 28 different genomic fragments coding for MyoHC isoforms have been isolated (Gerlach *et al.*, 1990).

The polymorphism of the myosin is also enhanced by the variety of the multiple myosin light chains. Like in mammals, the fast light chain LC1f, LC2f and LC3f are expressed in the fast-twitch fibres in fish and the slow light chain LC1s and LC2s are expressed in slow-twitch fibres. In salmonids, a duplication of the light chain LC1f has been shown by SDS–PAGE (Martinez *et al.*, 1991, 1993). In Arctic charr the fastest migrating LC1f isoform seemed to be predominant during embryonic and euleutheroembryonic periods compared to the slowest migrating LC1f isoform which is predominant in the 5-year old fish (Martinez *et al.*, 1991). According to fish strain in *Salvelinus* a third LC1f has been detected (Martinez & Christiansen, 1994). However, during embryogenesis in Atlantic salmon, no developmental light chain isoform, like the mammalian LC1emb (Whalen, Butler-Browne & Gros, 1978) has been observed (Perzanowska, 1979). Only the ratio LC1/LC3 decreases during development. In higher vertebrates the myosin light chains constitute a multigene family, however, the alkali light chain LC1f and LC3f or LC1emb and LC1A result from an alternative splicing of a single gene. Up to the present, in fish, no characterization of the light chain genes has been reported. Nevertheless, in grey mullet, the determination of amino acid sequence of alkali light chains revealed that it originates from two different genes (Dalla Libera *et al.*, 1991). Except for a polymorphism of the light chains due to fibre types, no developmental light chain isoform has been observed at the protein level (Fauconneau, personal communication). Thus, we have preferred to study the polymorphism of the MyoHC as marker of maturation and muscle growth in trout.

Contrarily to what occurs in mammals or birds, in fishes, little is known concerning the regulation of MyoHC expression by extrinsic factors (environment, feeding) or intrinsic factors (hormones). The study of muscle characteristics are modified according to the environmental temperature in which the fish live. In response to temperature acclimation, contractile activities are modified with changes in the ATPase activities. This is the physiological result of the switches in gene expression of the MyoHC (Goldspink, 1995). Different subfragments-1 have also been characterized in carp acclimated to warm or cold environmental temperature (Watabe, Guo & Hwang, 1994), as well as differences in the number and the primary structure of the 3′ part of the rod corresponding to the light meromyosin have been observed between carp acclimated to 10 °C or 30 °C (Watabe *et al.*, 1995). In this species, one MyoHC isoform gene is expressed in small white muscle fibres of adult warm acclimated fish (Ennion *et al.*, 1995).

Concerning studies of the MyoHC regulation, a decrease of MyoHC content has been observed after an estradiol administration in salmonids (Nazar *et al.*, 1991) but it is not known if it is a specific decrease or a general decrease of all contractile proteins. The variation of the food ration size in cod has induced changes of the MyoHC and MyoHC mRNA contents in response to fasting. However, particular modifications of specific MyoHC expression related to muscle growth process could not be detected without specific MyoHC probes. Thus, we have started to investigate the polymorphism of MyoHC mRNAs within the myotomal muscles from rainbow trout by obtaining specific MyoHC cDNA probes.

Myosin isoforms and subunits in skeletal muscle

In fish muscle, like in mammals or birds, myosin polymorphism has been studied by a variety of electrophoretic techniques, in both non-dissociating or dissociating conditions. However, the use of these techniques is more difficult for fish myosin which are more thermo-unstable than mammalian myosin (Connell, 1961; Ogawa *et al.*, 1993). This explains why there is scarce information concerning myosin polymorphism during development with respect to fibre types. In crucian carp, tench, pond loach and perch, three native myosins (FM1, FM2, FM3) were found in white muscle. A fourth isoform was detected in the carp, goldfish and stone perch migrating faster than the FM1 (Karasinski, 1993). Four native myosins were found in the white muscle of mackerel, eel, roach, cod, blue whiting, Norway haddock and spotted wolf-fish (Martinez *et al.*, 1989; Chanoine *et al.*, 1990; Karasinski & Kilarski, 1989; Martinez, Ofstad & Olsen, 1990*a*). However, the predominant isoforms were different according to species. For red muscle a higher diversity in the number of myosin and their mobility amongst fish species has been observed. One isoform was detected in the red myotomal muscle of carp and goldfish whereas two myosins were separated in mackerel, eel, Arctic charr, tench, pond loach, crucian carp (Martinez *et al.*, 1989, 1991; Chanoine *et al.*, 1990; Karasinski, 1993; Karasinski, Zawadowska & Supikova, 1994) and at least four myosins in perch and stone perch (Karasinski, 1993).

This polymorphism has been studied at the myosin subunit level by electrophoresis under dissociating conditions. It is well known that part of the polymorphism is related to regulatory light chains (LC1 and LC3) as previously described in rainbow trout by Perzanowska, Gerday & Focant (1978) and Rowlerson *et al.* (1985) and the myosin heavy chain (MyoHC).

In rainbow trout, we have shown only one separated band corresponding to MyoHC in the white muscle which had an electrophoretic mobility lower than the MyoHC detected in the red muscle by electrophoresis under dissociating conditions (Fig. 1). This result corroborates those already obtained in salmonids (rainbow trout) by Karasinski

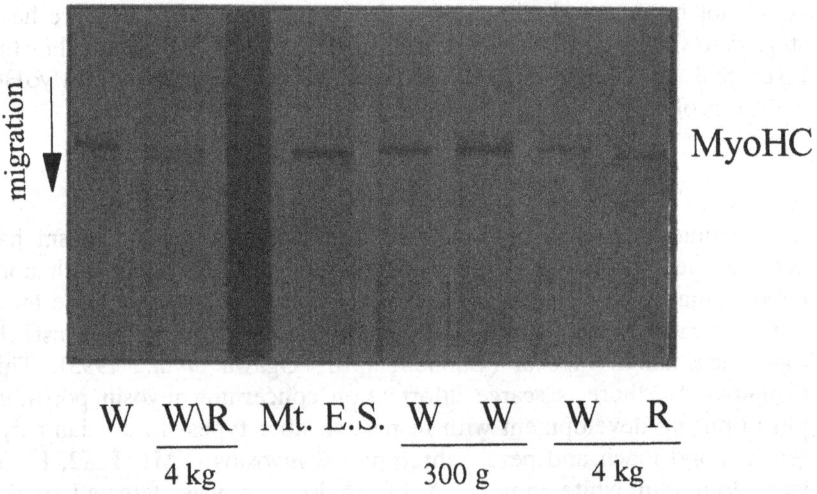

Fig. 1. SDS–PAGE of the myosin heavy chain from embryo at eyed stage (E.S.) or white (W) and red (R) myotomal muscle of rainbow trout during development and from myotube (Mt.) derived from cultivated trout primary satellite cell culture.

(1993), (rainbow trout and brown trout) by Fauconneau *et al.* (1994) and (Arctic charr, Atlantic salmon) by Martinez *et al.* (1991, 1993). These results have also been observed in other fish species (mackerel, sea bass, barbel, crucian carp, roach and tench) (Martinez *et al.*, 1989; Bassani & Dalla Libera, 1991; Huriaux, Vandewalle & Focant, 1991; Karasinski, 1993). In carp, goldfish and pond loach, the co-migration of MyoHCs from white and red muscle are generally observed. For some species, more than one MyoHC has been observed within the same muscle. Up to three MyoHCs have been separated in red muscle of eel (Chanoine *et al.*, 1990) and two MyoHCs in the white muscle of herring and stone perch (Martinez, Ofstad & Olsen, 1990*b*; Karasinski, 1993).

During development of rainbow trout, similar to what occurs in other salmonids, we have observed only one band in the white skeletal muscle from embryos (eyed stage) to adult (Fig. 1). However, during ontogenesis in Arctic charr, a second minor faster migrating band has been observed which is present only in Arctic charr embryo (Martinez *et al.*, 1991). From myotubes derived from trout primary satellite cell culture, one band was exhibited with a lower electrophoretic mobility than the band observed in white muscle and in embryos at eyed stage (Fig. 1). At this stage, embryo trunk muscle is also composed of myoblasts and myotubes. This isoform could correspond to a single early MyoHC which is also already expressed in embryos but is not discriminated by SDS–PAGE, due to comigration with other isoforms, or to an early MyoHC, which is not expressed any more. Another hypothesis is the expression in the *in vitro* myogenesis system of a modified MyoHC migrating very differently from the MyoHC expressed in the red muscle (slow). To confirm these hypothesis, specific trout antibodies against MyoHC isoforms and Western blot analysis are required.

Furthermore, different electrophoretic patterns are observed for the same species by modification of the pH or glycerol concentration (Huriaux *et al.*, 1991) and the difficulties to separate MyoHC isoforms using electrophoresis under denaturing conditions or to discriminate isoforms by immunohistochemical studies in salmonids (Rowlerson *et al.*, 1985) have required other techniques such as the peptide mapping to investigate further the MyoHC polymorphism according to their sensitivity to proteolytic digestions. Several transitory MyoHC isoforms have been detected within the white myotomal muscle during the development by Martinez *et al.* (1991, 1993). In Arctic charr, the sequential expression of at least six MyoHCs has been displayed in the developing white muscle using this technique and in Atlantic salmon two MyoHCs have been displayed; one MyoHC referred as neonatal and expressed in white muscle of parrs which is still detectable in adult fish and another isoform an adult fast type II-2 MyoHC which emerges several months after the smoltification (Martinez *et al.*, 1993). Rather than using this technique which allows the comparative characterization of the main MyoHC expressed at different stages but for which the interpretation of the results are difficult, we have preferred to analyse the polymorphism of the MyoHC at the mRNA level using specific isoform MyoHC cDNA probes.

Polymorphism of the myosin heavy chain mRNA throughout development

The transitory expression of MyoHCs within the same muscle or the expression of fibre-type specific MyoHC isoforms has been well characterized in mammals and birds. To investigate the characterization of the MyoHC isoforms we have created two trout cDNA libraries at two different developmental stages. One library (A) has been created from mRNA isolated from the trunk skeletal muscle of embryos at the eyed stage. According to the results from Nag & Nursal (1972), at this stage in rainbow trout, trunk muscle is composed of myoblasts and myotubes. The second library (B) has been produced using mRNA extracted from the white myotomal muscle of sub-adult rainbow trout of 300 g. Muscle growth at this stage results from an equal contribution of hyperplasia and hypertrophy. The libraries were screened using an heterologous fast skeletal MyoHC cDNA probe of adult mouse (kindly provided by M.E. Buckingham).

Three clones have been more precisely studied (Fig. 2). They encode the 3′ part of the MyoHC including the untranslated region. The nucleotide sequence analysis demonstrated that the clone A1 (isolated from the embyro library) and the clone B6 (isolated from the sub-adult library) encode for the same MyoHC isoform. The clone B8 sharing 98% identity with the clone B6 in the coding and uncoding region has been considered as an allelic isoform because of the absence of insertion or deleted sequence, and the absence of a variable region in the specific MyoHC isoform untranslated region. This MyoHC transcript, expressed strongly in white skeletal muscle, is a fast skeletal MyoHC (Fig. 3). The faint expression observed in red muscle could be due to the presence of fast skeletal MyoHC within this muscle demonstrated by immunohistochemical studies in trout (Rowlerson *et al.*, 1985). The most interesting result was the continuous expression of this MyoHC isoform throughout development. A cDNA probe corresponding to the

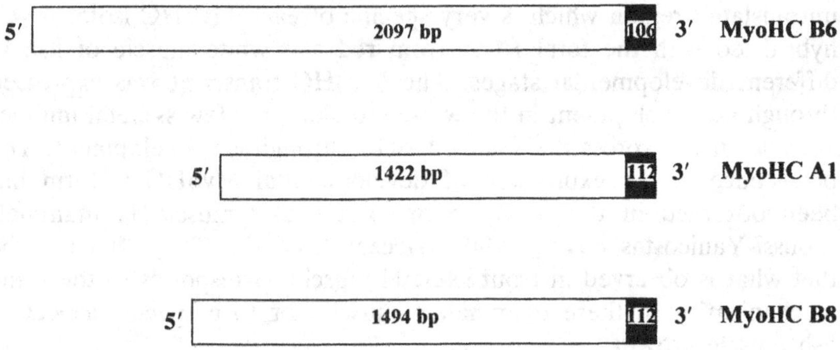

Fig. 2. Alignment of trout cDNAs coding for the 3′ part of the myosin heavy chain and the untranslated region. Clone A1 has been isolated from the embryo cDNA library and clone B6 and B8 from a sub-adult white muscle cDNA library. □ coding region, ■ uncoding region

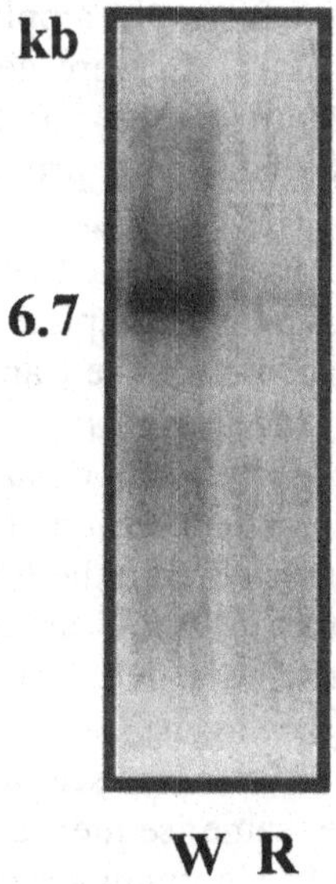

Fig. 3. Northern blot analysis of mRNA extracted from white (W) and red (R) myotomal muscle of trout using the 3′ untranslated region of the B6 MyoHC cDNA clone.

untranslated region which is very specific of each MyoHC isoform was hybridized with the total RNA from red and white muscle of fish at different developmental stages. The MyoHC transcript was expressed throughout development in the white muscle. Only few skeletal muscles are known to express the same MyoHC throughout development. The persistence in the expression of developmental MyoHC isoform has been observed in the masseter or extra-ocular muscle in mammals (Soussi-Yanicostas *et al.*, 1990; Wieczorek *et al.*, 1985). It could be that what is observed in trout skeletal muscle corresponds to the same mechanism as in these mammalian muscles or to a specific aspect of fish muscle growth.

Another method used to analyse the polymorphism of the MyoHC at the mRNA level is the 3′ RACE PCR technique (rapid amplification of the 3′ cDNA ends by polymerase chain reaction). This technique allows the amplification of PCR products from mRNA template which little sequence information is available and which contains the 3′ untranslated region. We have synthesized first strand cDNA initiated at the poly (A) tail of mRNA using an adapter primer (AP) annealed to an oligo (dT) provided by 3′ RACE system for rapid amplification of cDNA ends (Gibco BRL). PCR products of 1.2 kb were generated by amplification of the 3′ part from trout MyoHC isoforms and the variable untranslated region using a trout MyoHC sense primer from the coding region (5′-GACTCCATGCAGAGCAC-3′) and an UAP antisense primer corresponding to a portion identical to that of the adapter primer. The PCR were performed using, as template for the first strand cDNA synthesis, total RNA extracted from red and white myotomal muscle at different developmental stages from myotubes up to adult trout muscle; 35 cycles were performed with 1 min denaturation at 95 °C, 1 min annealing at 50.5 °C and 2 min extension at 72 °C. The PCR fragments were digested with *Pst*I and *Spe*I and the linearized *Pst*I-*Spe*I fragment of 0.3 kb, corresponding to the last 65 amino acids and the untranslated region characteristic of each MyoHC isoform was subcloned into bluescript II KS^+ phagemid (Stratagene).

One main PCR product was isolated from different muscle at different developmental stages. It codes for the same sequence as the overlapping sequence of the MyoHC clone B6. Contamination by the MyoHC clone B6 could be excluded because the PCR products contained the sequence of the adapter primer provided by the 3′ RACE PCR kit and used to synthesize the first strand cDNA.

Two other MyoHC PCR products were isolated from muscle at hatching (P24) or from trunk muscle of juvenile fish (85g) (P41). From the comparison of the coding sequence of the clone P24 with the clone B6, eight amino acids were different with four amino acids in the last 10 amino acids (Fig. 4). This difference results from a deletion of three nucleotides (TTG) located 11 nucleotides upstream to the end of the coding region. For the clone P41, two amino acids are divergent on the 65 amino acids overlapping with the clone B6. The comparison of nucleotide sequences of the untranslated region for these clones has shown a high identity with four or three divergent nucleotides for clones P24 and P41 with a deleted sequence of 17 nucleotides and an insertion

```
                                    10
Leu Gln Met Lys Val Lys Ala Tyr Lys Arg His Ser Glu Glu Ala Glu Glu Ala  B6
--- --- --- --- --- --- --- --- --- --- Gln Ala --- --- --- --- --- ---  P24
--- --- --- --- --- --- --- --- --- --- --- --- --- --- --- --- --- ---  P41

    20                                      30
Ala Asn Gln His Met Ser Lys Phe Arg Lys Val Gln His Glu Leu Glu Glu Ala  B6
Ser --- --- --- --- --- --- --- --- --- --- --- Asn --- --- --- --- ---  P24
--- --- --- --- --- --- --- --- --- --- --- --- --- --- --- --- --- ---  P41

            40                                          50
Glu Glu Arg Ala Asp Ile Ala Glu Thr Gln Val Asn Lys Leu Arg Ala Lys Thr  B6
--- --- --- --- --- --- --- --- --- --- --- --- --- --- --- --- --- ---  P24
--- --- --- --- --- --- --- --- Ser --- --- --- --- --- --- --- --- Ala  P41

                    60
Arg Asp Ser Gly Lys Gly Lys Glu Val Ala Glu  *                 B6
--- --- --- --- --- --- Asn Gln Ala Glu *                      P24
--- --- --- --- --- --- --- --- --- --- ---  *                 P41


ACAAGACCAA AGTATT     GAAGATCAA AGTCTTACCA TTTTCCTGTG TTGCATAAAA B6
---------- ------                        -- ---------- --T------- P24
----A----- ---C--ACCAT                   -- -----A---- ---------- P41

TATGATTTTC ATGGTGAAAT GTTGAGCATT GATTAAAAAC ATGTGGATC TA         B6
---------- ---------- ---------- ---------- ----C-GC- --CATACA   P24
---------- ---------- ---------- ---------- --------- --CATACA   P41
```

Fig. 4. Alignment of the amino acid sequence and nucleotide sequence of the untranslated region from different MyoHC PCR products.

of five nucleotides for the clone P41 only. Furthermore, a 1.4 kb MyoHC cDNA, isolated from the trout embryo cDNA library, shares a total identity with the overlapping nucleotide sequence of the clone P24.

Thus, the three different clones obtained showed a weak nucleotide sequence variation in the coding region and the 3′ untranslated region. This could be explained by two hypotheses. The diversification of highly conserved genes could be related to gene duplication resulting from the tetraploidization of the genome occurring in fishes. For duplicated genes, deleted or inserted sequences in the untranslated region such as those observed between the clones B6, P24 and P41 have already been reported. The differential behaviour of the duplicated gene could lead to an evolutionary modification of the coding sequence optimizing it to a different function to that of the ancestral gene or to changes in gene regulation to allow deployment at a new site or new time. The second hypothesis is that the clone P24 and the clone isolated from the muscle embryo cDNA library correspond to a different MyoHC isoform as those previously described with amino acid differences within the last 10–15 amino acids as it is classically observed between MyoHC isoforms (Radice & Malacinski, 1989). Further investigations are required to test such hypotheses. In carp, different MyoHC isoforms have been characterized, corresponding to specific fibre-type or developmental MyoHC isoforms using a similar approach (for review see Ennion *et al.*, this volume).

We are carrying out further research of the MyoHC cDNAs belonging to specific isoforms which are involved either in the muscle growth process or related to an environmental adaptation.

Regulation of MyoHC mRNA expression

The synthesis of muscle protein and especially of myosin is controlled by intrinsic factors and environmental factors such as temperature. The response to these different factors could be separated as non-specific changes related to growth and specific changes related to maintenance of functionality. In fish, growth includes *de novo* synthesis of contractile proteins due to hyperplasic growth of muscle and synthesis of contractile protein equipment allowing a functional adaptation to a specific (thermal) environment. The mechanisms by which different factors affect synthesis of contractile protein is not yet known in fish. It could be emphasized for data obtained at whole protein level that short-term response results from changes in gene expression (Houlihan, 1991). However, little if any information is available in fish on the regulation of specific protein synthesis and less on the regulation of MyoHC synthesis.

The effect of growth hormone could be used as an illustration of the possible regulation of MyoHC by different factors. It is well known that growth hormone could stimulate amino acid and glucose absorption in the cell and especially in muscle as a short-term effect and whole protein synthesis machinery as a long-term effect. In fish, short-term stimulation (a few hours) of amino acid uptake (Skyrud *et al.*, 1989) and protein synthesis (Matty, 1986) by growth hormone has been reported. Long-term (a few weeks) administration of growth hormone results in stimulation of RNA synthesis with few changes in efficiency of protein synthesis (Foster *et al.*, 1991). At cellular levels, these effects are associated with an increase in hyperplasic growth of muscle (Maisse *et al.*, 1993). Thus it is especially interesting to analyse the short-term and medium-term effect of growth hormone on the expression of a specific contractile protein such as MyoHC.

In a first experiment, fish (150 g mean body weight) were injected with a single dose (0.5 μg/g body weight) of rainbow trout recombinant growth hormone (rtGH). Then during a 24-hour period, RNA/protein ratio and MyoHC mRNA (analysed by slot–blot and Northern blot with the B6 cDNA probes) were measured in muscle. Although post-injection changes in MyoHC mRNA seem to have occurred earlier in the GH group than those in the control group, no significant differences were observed for the whole 24-hour period between the two groups in RNA/protein ratio and in relative (/μg total RNA) and absolute (/g protein) MyoHC mRNA (Fig. 5). There is no short-term change in MyoHC mRNA after a single injection of GH.

In a second experiment, fish (300 g mean body weight) were injected three times at 12-hour intervals with rtGH (1 μg/g body weight). Then fish were sampled either 6 hours and 24 hours after the last injection and the same parameters were measured (Fig. 6). No significant differences in RNA/protein ratio were observed suggesting that overall protein synthesis machinery is not yet affected. Both relative and absolute MyoHC mRNA levels were significantly higher at 6 hours after the injection of rtGH as compared to the control fish. Thus rtGH is able to stimulate MyoHC mRNA synthesis after a period of time compatible with the implication of IGF. In this experiment it has been demonstrated that expression of IGFs by liver is significantly stimulated (Le Gac, personal communication). The response of MyoHC mRNA is observed, however, with rather high dose and only temporarily (no significant differences in the same parameters were observed 24 hours after the last injection).

Thus in the long-term response of muscle protein synthesis to growth hormone, a relative change in the expression of some specific mRNAs

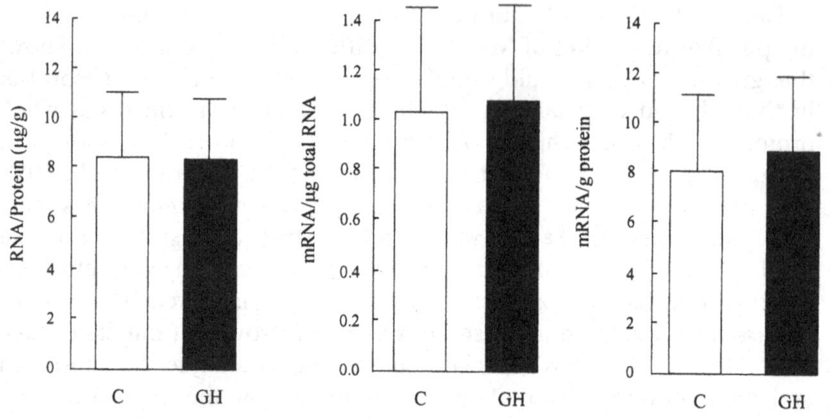

Fig. 5. Effect of a single injection of recombinant trout growth hormone on RNA/protein ratio and mRNA of MyoHC in muscle of rainbow trout. The results are the mean values ±SEM of 42 fish.

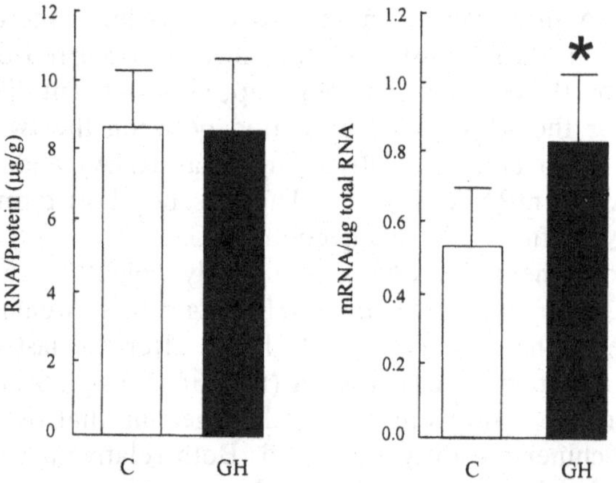

Fig. 6. Effect of a repeated injection of recombinant trout growth hormone on RNA/protein ratio and mRNA of MyoHC in muscle of rainbow trout. The results are the mean values ±SEM of 11 fish.

such as those of MyoHC have been observed but only temporarily. A decrease of MyoHC and relative (calculated per g wet weight) MyoHC mRNA contents have been observed, for instance, during the response to fasting in cod (Von der Decken & Lied, 1989). It could be suggested that, in most of the cases there is a coordinated stimulation of the different contractile proteins in response to factors affecting growth.

From the data obtained in these two experiments, we could also conclude that the MyoHC mRNA level is relatively stable. This is especially important for the maintenance of muscle activity.

Conclusion

Whole myosin polymorphism observed in salmonids starts to be well characterized at the protein level by polymorphism of both MyoHCs and MyoLCs. However, the little differences between MyoHC isoforms are difficult to demonstrate at the protein level and require sequence analysis using molecular biology methods. We have obtained one MyoHC cDNA coding for the 3′ part of a fast skeletal myosin expressed throughout development in white myotomal muscle. From MyoHC isoforms characterized and expressed in the white skeletal muscle, the nucleotide and amino acid sequences are relatively conserved to indicate a duplicated gene or sequence coding for another MyoHC gene.

Acknowledgements

We are grateful to Dr P.Y. Rescan and Dr H. D'Cotta for many useful discussions and comments on the manuscript. This work was supported by a grant from INRA and the region of Brittany.

References

Bassani, V. & Dalla Libera, L. (1991). Myosin isoforms in white, red and pink muscles of the teleost sea bass (*Dicentrarchus labrax*, L.). *Basic and Applied Myology*, **1**, 153–6.

Buckingham, M.E. (1985). Actin and myosin multigene families: their expression during the formation of skeletal muscle. *Essays in Biochemistry*, **20**, 77–109.

Chanoine, C., Saadi, A., Guyot-Lenfant, M., Hebbrecht, C. & Gallien, Cl.L. (1990). Myosin structure in the eel (*Anguilla anguilla* L.). *FEBS Letters*, **277**, 200–4.

Connell, J.J. (1961). The relative stabilities of the skeletal muscle myosins of some animals. *Biochemical Journal*, **80**, 503–9.

Dalla Libera, L., Carpene, E., Theibert, J. & Collins, J.H. (1991). Fish myosin alkali light chains originate from two different genes. *Journal of Muscle Research and Cell Motility*, **12**, 366–71.

Ennion, S., Gauvry, L., Butterworth, P. & Goldspink, G. (1995). Small diameter white muscle fibres associated with growth hyperplasia in the carp (*Cyprinus carpio*) express a distinct myosin heavy chain gene. *Journal of Experimental Biology*. In press.

Fauconneau, B., Bonnet, S., Douirin, C., de Guilbert, B., Lefevre, F., Laroche, M. & Buvineau, C. (1994). Assessment of muscle biochemical and histochemical criteria for flesh quality in salmonids. In *Measures for Success*. ed. P. Kestemont, J. Muir, F. Sevila & P. Williot, pp. 225–238. CEMAGREF.

Foster, A.R., Houlihan, D.F., Gray, C., Medale, F., Fauconneau, B., Kaushik, S.J. & Lebail, P.Y. (1991). The effects of ovine growth hormone on protein turnover in rainbow trout. *General Comparative Endocrinology*, **81**, 111–20.

Gerlach, G., Turay, L., Malik, T.A., Lida, J., Scutt, A. & Goldspink, G. (1990). Mechanism of temperature acclimation in the carp: a molecular approach. *American Journal of Physiology*, **259**, R237–44.

Goldspink, G. (1995). Adaptation of fish to different environmental temperature by qualitative and quantitative changes in gene expression. *Journal of Thermal Biology*, **20**, 167–74.

Houlihan, D.F. (1991). Protein turnover in ectotherms and its relationships to energetics. In *Advances in Comparative and Environmental Physiology*, ed. R. Gilles, vol. 7, pp. 1–43. Berlin, Heidelberg: Springer-Verlag.

Huriaux, F., Vandewalle, P. & Focant, B. (1991). Myosin heavy chain isoforms in white, red and ventricle muscles of barbel (*Barbus barbus* L.). *Comparative Biochemistry and Physiology*, **100B**, 309–12.

Karasinski, J. (1993). Diversity of native myosin and myosin heavy chain in fish skeletal muscles. *Comparative Biochemistry and Physiology*, **106B**, 1041–7.

Karasinski, J. & Kilarski, W. (1989). Polymorphism of myosin isoenzymes and myosin heavy chains in histochemically typed skeletal muscles of the roach (*Rutilus rutilus* L., Cyprinidae, Fish). *Comparative Biochemistry and Physiology*, **92B**, 727–31.

Karasinski, J., Zawadowska, B. & Supikova, I. (1994). Myosin isoforms in selected muscle fibre types of the pond loach *Misgurnus fossilis* L. *Comparative Biochemistry and Physiology*, **107B**, 249–53.

Maisse, G., Bonnieux, F., Fauconneau, B., Faure, A., Cloaguen, Y., Lebail, P.Y., Prunet, P. & Rainelli, P. (1993). The zootechnical and socio economical potential impacts of the use of rtGH in salmon farming. *INRA Productions Animales*, **6**, 167–83.

Martinez, I., Olsen, R.L., Ofstad, R., Janmot, C. & d'Albis, A. (1989). Myosin isoforms in mackerel (*Scomber scombrus*) red and white muscle. *FEBS Letters*, **252**, 69–72.

Martinez, I., Ofstad, R. & Olsen, R.L. (1990*a*). Electrophoretic study of myosin isoforms in white muscles of some teleost fishes. *Comparative Biochemistry and Physiology*, **96B**, 221–7.

Martinez, I., Ofstad, R. & Olsen, R.L. (1990*b*). Intraspecific myosin light chain polymorphism in the white muscle of herring (*Clupea harengus harengus*, L.). *FEBS Letters*, **265**, 23–6.

Martinez, I., Christiansen, J.S., Ofstad, R. & Olsen, R.L. (1991). Comparison of myosin isoenzymes present in skeletal and cardiac muscles of the Arctic charr *Salvelinus alpinus* (L.). Sequential expression of different myosin heavy chains during development of the fast white skeletal muscle. *European Journal of Biochemistry*, **195**, 743–53.

Martinez, I., Bang, B., Hatlen, B. & Blix, P. (1993). Myofibrillar proteins in skeletal muscles of parr, smolt and adult Atlantic salmon (*Salmo salar* L.). Comparison with another salmonid, the Arctic charr *Salvelinus alpinus* (L.). *Comparative Biochemistry and Physiology*, **106B**, 1021–8.

Martinez, I. & Christiansen, J.S. (1994). Myofibrillar proteins in developing white muscle of the Arctic charr, *Salvelinus alpinus* (L.). *Comparative Biochemistry and Physiology*, **107B**, 11–20.

Matty, A.J. (1986). Nutrition, hormones and growth. *Fish Physiology and Biochemistry*, **2**, 141–50.

Nag, A.C. & Nursall, J.R. (1972). Histogenesis of white and red muscle fibres of trunk muscles of a fish *Salmo gairdneri. Cytobis*, **6**, 227–46.

Nazar, D.S., Persson, G., Olin, T., Waters, S. & von der Decken, A. (1991). Sarcoplasmic and myofibrillar proteins in white trunk muscle of salmon (*Salmo salar*) after estradiol treatment. *Comparative Biochemistry and Physiology*, **98B**, 109–14.

Ogawa, M., Ehara, T., Tamiya, T. & Tsuchiya, T. (1993). Thermal stability of fish myosin. *Comparative Biochemistry and Physiology*, **106B**, 517–521.

Pedrosa, F., Butler-Browne, G.S., Dhoot, G.K., Fischman, D.A. & Thornell, L.E. (1989). Diversity in expression of myosin heavy chain isoforms and M-band proteins in rat muscle spindles. *Histochemistry*, **92**, 185–94.

Perzanowska, A., Gerday, C. & Focant, B. (1978). Light chains of trout myosin isolation and characterization. *Comparative Biochemistry and Physiology*, **60B**, 295–301.

Perzanowska, A. (1979). The light chain composition of embryonic myosin. *Comparative Biochemistry and Physiology*, **63B**, 189–92.

Radice, G.P. & Malacinski, G.M. (1989). Expression of myosin heavy chain transcripts during *Xenopus laevis* development. *Developmental Biology*, **133**, 562–8.

Rowlerson, A., Scapolo, P.A., Mascarello, F., Carpene, E. & Veggetti, A. (1985). Comparative study of myosins present in the lateral muscle of some fish: species variations in myosin isoforms and their distribution in red, pink and white muscle. *Journal of Muscle Research and Cell Mobility*, **6**, 601–40.

Sartore, S., Mascarello, R., Rowlerson, A., Gorza, L., Ausoni, S., Vianello, M. & Shiaffino, S. (1987). Fibre types in extraocular muscles: a new myosin isoform in the fast fibres. *Journal of Muscle Research and Cell Mobility*, **8**, 161–72.

Skyrud, T., Anderson, O., Aleström, P. & Gautvik, K.M. (1989). Effects of recombinant human growth hormone and insulin-like growth factor 1 on body growth and blood metabolites in brook trout (*Salvelinus fontinalis*). *General Comparative Endocrinology*, **75**, 247–55.

Soussi-Yanicostas, N., Barbet, J.P., Laurent-Winter, C., Barton, P. & Butler-Browne, G.S. (1990). Transition of myosin isozymes during development of human masseter muscle. Persistence of developmental isoforms during postnatal stage. *Development*, **108**, 239–49.

Staron, R.S. & Johnson, P. (1993). Myosin polymorphism and differential expression in adult human skeletal muscle. *Comparative Biochemistry and Physiology*, **106B**, 463–75.

Termin, A., Staron, R.S. & Pette, D. (1989). Changes in myosin heavy chain isoforms during chronic low-frequency stimulation of rat fast hindlimb muscles. *European Journal of Biochemistry*, **186**, 749–54.

Von der Decken, A. & Lied, E. (1989). Myosin heavy chain synthesis in white trunk muscle of cod (*Gadus morhua*) fed different ration sizes. *Fish Physiology and Biochemistry*, **6**, 333–40.

Watabe, S., Guo, X.F. & Hwang, G.C. (1994). Carp express specific isoforms of the myosin cross-bridge head, subfragment-1, in association with cold and warm temperature acclimation. *Journal of Thermal Biology*, **19**, 261–8.

Watabe, S., Imai, J.I., Nakaya, M., Hirayama, Y., Okamoto, Y., Masaki, H., Uozumi, T., Hirono, I. & Aoki, T. (1995). Temperature acclimation induces light meromyosin isoforms with different primary structures in carp fast skeletal muscle. *Biochemical and Biophysical Research Communications*, **208**, 118–25.

Whalen, R.G., Butler-Browne, G.S. & Gros, F. (1978). Identification of a novel form of myosin light chain present in embryonic muscle tissue and cultured muscle cells. *Journal of Molecular Biology*, **126**, 415–31.

Whalen, R.G., Sell, S.M., Butler-Browne, G.S., Schwartz, K., Bouveret, P. & Pinset-Häström, I. (1981). Three myosin heavy-chain isozymes appear sequentially in rat muscle development. *Nature*, **292**, 805–9.

Wieczorek, D.F., Periasamy, M., Butler-Browne, G.S., Whalen, R.G. & Nadal-Ginard, B. (1985). Co-expression of multiple myosin heavy chain genes, in addition to a tissue-specific one, in extraocular masculature. *Journal of Cell Biology*, **101**, 618–29.

N. MACLEAN, M.S. ALAM, A. IYENGAR and A. POPPLEWELL

Transient expression of reporter genes in fish as a measure of promoter efficiency

Introduction

One of the problems associated with the attempted induction of transgenesis in fish is the high incidence of transient expression (see discussion in Iyengar, Müller & Maclean, 1996). This phenomenon is troublesome if one is attempting to correlate transgene integration into chromosomal DNA with transgene expression, at least in early developmental stages, since unintegrated copies of the transgenes replicate and express. The problem is compounded by the fact that many of the transiently expressing individuals never incorporate transgene copies and therefore never become mature transgenic animals. However, transient transgene expression can be used to determine the activity of heterologous gene regulatory regions if these are spliced to effective reporter genes.

In the study presented here, two equivalent *lacZ* containing constructs driven, respectively, by carp and rat β-actin regulatory sequences, were introduced into fertilized eggs of tilapia (*Oreochromis niloticus*) or rainbow trout (*Oncorhynchus mykiss*) and the levels of transgene expression measured in homogenates of embryos. In addition, X-gal staining was used to check visually on the status of expressing embryos and Southern blotting to indicate the conformation of the transgene in the young fish.

Several fish species have been used as model systems in which to test the activity of regulatory sequences *in vivo* (Stuart, McMurray & Westerfield, 1988; Chong & Vielkind, 1989; Winkler, Vielkind & Schartl, 1991; Winkler *et al.*, 1992; Moav *et al.*, 1993; Gong & Hew, 1991), but no extensive comparison on the efficiency of piscine and non-piscine sequences has been previously reported.

The data presented indicate that transient expression can be used to monitor the efficiency of 5′ regulatory regions in driving transgene expression, and that different regulatory regions express differentially. Promoterless *lacZ* constructs have also been used to indicate that the expression observed is indeed a result of the activity of the 5′ regulatory regions.

Comparison of *lacZ* expression in fish embryo homogenates

In this study, two transgene constructs were used, with the aim of comparing, as near as possible, equivalent regulatory regions from rat and carp β-actin genes (see Figs. 1 and 2). The β-actin gene was selected, partly because it is ubiquitously expressed, but also because genes had been cloned from fish and mammalian sources which possessed substantial 5′ regions upstream of the coding sequence and therefore afforded good examples for comparison. The two sequences of the β-actin gene are similar in terms of size, the locations of the exon/intron junctions and the possession of an untranslated first exon (Liu *et al.*, 1990*a,b*). The conserved sequences in each include CCAAT and TATA motifs and a $CC(A/T)_6GG$ sequence (CArG) motif (Nudel *et al.*, 1983; Liu *et al.*, 1991).

pP3PA-*lacZ*-CarpβA contains the carp β-actin promoter with 3.3 kb of 5′ flanking sequence, an untranslated exon 1, intron 1, and the first five bases of exon 2, fused at the ATG start codon to the *lacZ* reporter gene and the SV40 poly A signal. It was assembled by inserting a 4.5 kb fragment from a SalI/NcoI (partial) digest of pCβA4.6 into the Xho/NcoI site of pP3PAlacZ(A. Popplewell & S. Morley, Edinburgh). The original carp β-actin sequence was provided by Professor Perry Hackett in Minnesota. Plasmid DNA was restricted with *Spe*I and an approximately 7.9 kb fragment of the insert DNA, CarpβAlacZ, was recovered and used for microinjection.

The pP3PAlacZ–RatβA contains a rat β-actin gene promoter (Nudel *et al.*, 1983) with 3.4 kb 5′ flanking sequences, an untranslated exon 1, intron 1 and the first nine bases of exon 2, fused at the ATG start codon to the *lacZ* reporter gene and the SV40 poly A signal. It was prepared by inserting a 4.6 kb fragment of an EcoRI/NcoI digest (partial) of pRβA8.0 into the *Mun*I/*Nco*I site of pP3PAlacZ (A. Popplewell & S. Morley, Edinburgh). The original rat β-actin sequence was provided by D. Yaffe of the Weizmann Institute of Science, Rehovot, Israel. The plasmid DNA was restricted with the enzyme SpeI and an approximately 7.9 kb fragment of the insert DNA, RatβA-lacZ, was recovered and used for microinjection. Microinjection of the linear transgene fragments was carried out by introducing approximately 2×10^5 copies of the DNA into the germinal disc of fertilized tilapia eggs as described by Rahman & Maclean (1992) or by introducing

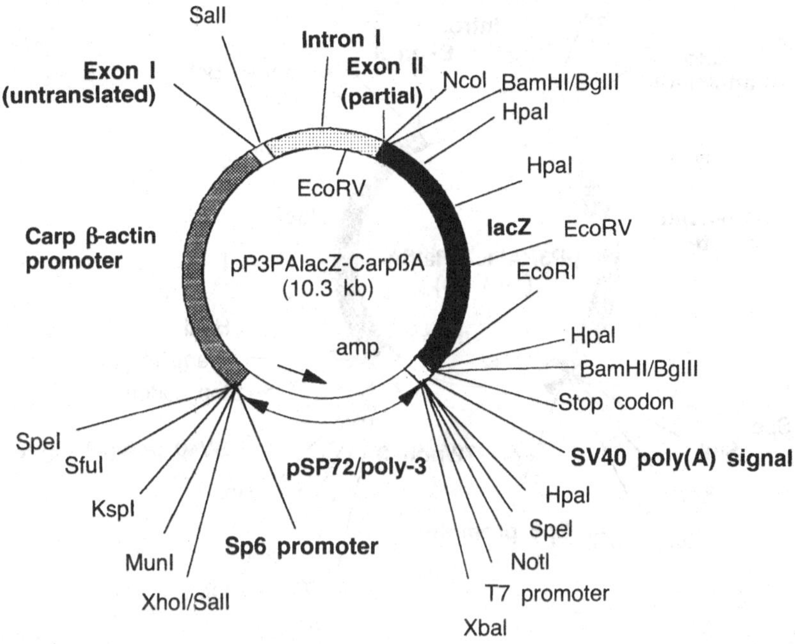

(*a*)

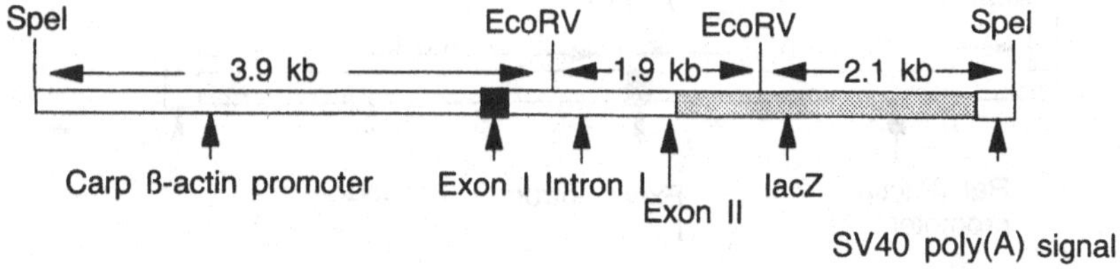

(*b*)

Fig. 1. (*a*). sp72 plasmid (Promega) containing the *E.coli lacZ* gene spliced to 4.6 kb regulatory sequences of the carp β-actin gene and SV40 poly A signal. (*b*). 7.9 kb *SpeI/SpeI* linear fragment of the CarpβAlacZ gene construct.

approximately 10^6 copies of the DNA into the cytoplasm of fertilized trout eggs as described by Penman *et al.* (1991). In the case of tilapia, embryos/fry were analysed on days between 1 and 8 after fertilization. Assay for *lacZ* expression involved homogenizing the embryos and using

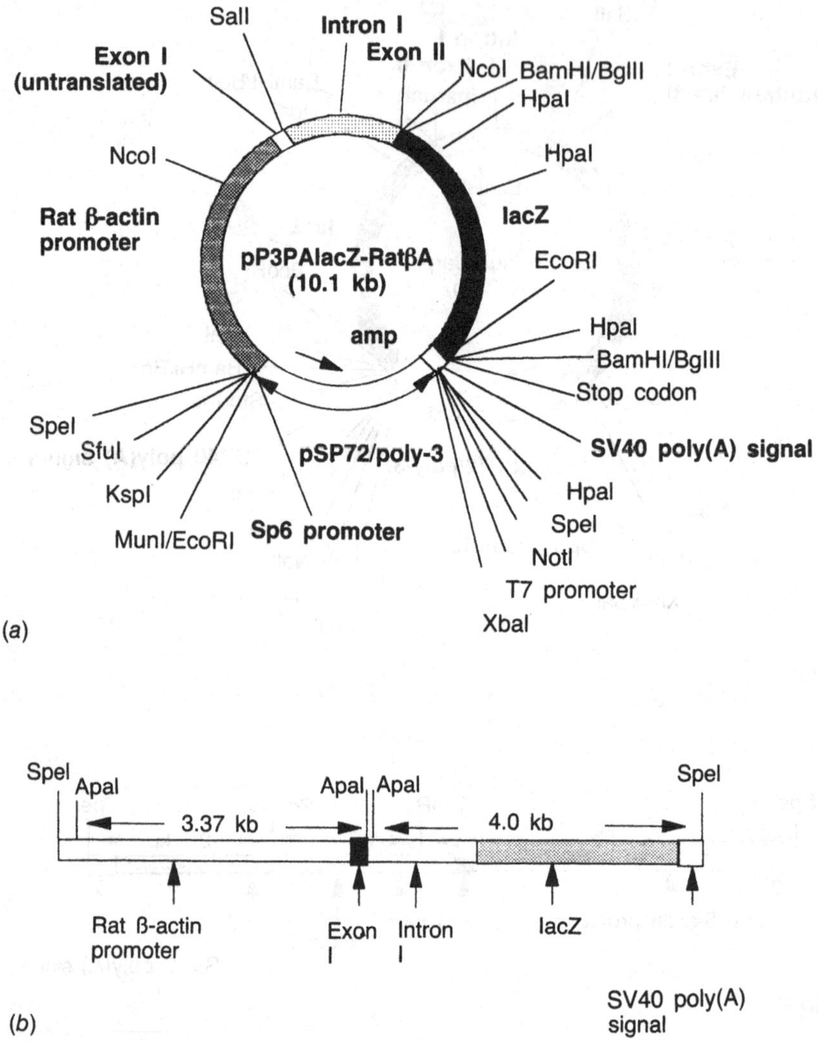

Fig. 2. (*a*). sp72 plasmid (Promega) containing the *E.coli lacZ* gene spliced to 4.6 kb regulatory sequences of the rat β-actin gene and SV40 poly (A) signal. (*b*). 7.9 kb *Spe*I/*Spe*I linear fragment of the RatβAlacZ gene construct.

the MUG (methyl umbelliferyl-β-D-galactoside) substrate according to the method of Braell (1991). Some embryos were also stained for *in situ* expression using X-gal (5-bromo-4-chloro-3-indolyl-β-D-galactoside) as a substrate. Trout were assayed only by MUG assay, either on day 7 or at days 50–55 after fertilization when the alevins are free swimming.

Some of the embryos used for MUG assays were also used for DNA extraction and Southern blotting. This involved splitting the homogenate of an embryo into two equal aliquots and using one half for MUG assay and one half for DNA extraction. DNA purification followed by slot blotting and Southern blotting were carried out as previously described (Penman *et al.*, 1991).

In addition to analyses of embryos and fry grown from injected embryos, control animals were also assayed, being those that had grown from equivalent batches of non-injected eggs.

As seen in Table 1, the levels of expression in the injected embryos exceeded the control levels in many of the embryos, and the data can be best summarized in the following series of observations.

1 In tilapia, more than 50% of the injected embryos expressed at levels above controls when the carpβAlacZ construct was injected, while the incidence with the ratβAlacZ construct was substantially less.
2 The control levels of expression with tilapia increased with the stage of development, and so, no doubt as a result of increasing cell number, the level of expression needed to exceed control levels also increased.
3 The levels of expression with tilapia were variable within batches and between batches at different developmental stages, but at most stages the levels were much higher for the carp construct expression than for the rat construct.
4 There is a statistically significant difference in both incidence and levels of expression in tilapia at some, but not all, stages of development, when the carp and rat construct injected embryos are compared.
5 No convincing positives were recorded amongst control embryos nor amongst embryos injected with promoterless constructs, with either tilapia or trout.
6 The total number of trout eggs injected was greater than with tilapia, but the incidence of expression in trout was less. However, the construct differences recorded for tilapia in the observations listed in 2–5 above also apply to trout. A further presentation of the experimental data and the

Table 1. *Incidence and levels of expression of the two equivalent gene constructs (Carpβ AlacZ and Ratβ AlacZ) containing the carp and rat β-actin regulatory sequences*

(*a*) *Tilapia*

Developmental times	Incidence of expression (%)		Levels of expression	
	CarpβAlacZ	RatβAlacZ	CarpβAlacZ	RatβAlacZ
Day 1	46.2 (6/13)	30.0 (3/10)	R^a=167–1250 M=412	R^a=136–790 M=448
Day 2	57.1 (8/14)	37.5 (3/8)	R^a=145–6200 M=1620	R^a=154–6270 M=2208
Day 3	76.2 (16/21)	38.5 (5/13)	R^a=316–13 250 M=5375	R^b=210–1950 M=705
Day 4	58.3 (7/12)	33.3 (3/9)	R^a=300–31 000 M=9011	R^a=325–2240 M=1081
Day 5	75.0 (12/16)	23.8 (5/21)	R^a=1250–85 000 M=18 430	R^b=635–8800 M=2662
Day 6	88.9 (8/9)	33.3 (3/9)	R^a=650–38 000 M=16 856	R^b=1100–2330 M=1790
Day 7	63.6 (7/11)	12.5 (1/8)	R=1055–10 300 M=3847	1030
Day 8	53.3 (8/15)	13.3 (2/15)	R^a=1645–109 000 M=31114	R^a=1850–4000 M=2925
Mean	64.8 ± 14.2	27.8 ± 10.2	11 060	1611

(*b*) *Trout*

Developmental stages	Incidence of expression		Levels of expression	
	Carp	Rat	Carp	Rat
Day 7 (gastrula)	4.9% (5/101)	3.9% (3/77)	R=2486–25 020 M=13 208	R=2426–3682 M=2953
Day 50–55	1.7% (2/121)	1.0% (1/107)	R=56 450–59 050 M=57 750	6600

R and M represent the range and mean of the levels of β-gal activity of the positive individuals in a group of embryos. Figures followed by different superscripts are significantly different ($P<0.05$). The incidence of expression has been compared by *t*-test and the levels of expression have been compared by Wilcoxon's rank test. No comparison test was possible at day 7 in the case of tilapia and at days 50–55 in the case of trout because only one fry was positive for expression of the rat gene construct. The number of non-injected controls assayed in each batch of tilapia was always equal to the number of injected individuals. In trout, however, the number of controls assayed ranged from 22 to 48.

Positively expressing fry were arbitrarily taken to be those expressing more than 120% of the highest control value at that stage. The highest control ranged from 60 at day 1 to 1200 at day 8 in tilapia.

tests for statistical significance will be found in another publication (Alam *et al.*, submitted).

Correlation between transgene presence and expression

When simultaneous MUG assays for β-gal expression and DNA analyses by Southern blotting were carried out in tilapia fry, every expressing individual was also found to contain copies of transgene sequences. However, two individuals which were found to be positive for the rat construct in Southern blots gave levels of β-gal expression no higher than controls. We interpret this to be further evidence for poor expression of the rat construct in tilapia. The banding patterns on the Southern blots indicated extensive transgene concatamerization, making it difficult to determine if there was evidence for chromosomal integration of transgene copies in any of the positive animals. Also, using the *in situ* X-gal staining procedure, which is less sensitive than the fluorometric MUG assay procedure, 30% (9/30) of the tilapia fry grown from carpβAlacZ-injected eggs demonstrated positive expression; however, none of the 30 fry analysed after ratβAlacZ injection had detectable β-gal expression. Again, we assume that this underlines the higher expression levels detectable in the carp construct-injected eggs as compared to those injected with the rat construct. With trout, only slot blots were carried out so there is no information about concatamerization or incorporation. However, four out of six embryos positive on slot blots for the carp construct did not express in terms of MUG assay, while six out of six positive on slot blots for the rat construct failed to express. Although these numbers are very low, they do emphasize that possession of transgene copies is itself insufficient to ensure expression, and that this is particularly true of the rat construct.

Why is expression level variable within and between constructs?

It is difficult to explain the variation in the expression observed within a batch of injected individuals but it may be attributed to several factors including transgene mosaicism and differential replication of the transgene DNA. This variation, however, appears to be a common feature with transgenic fish, and has been previously observed by a number of researchers (e.g. Stuart *et al.*, 1990; Winkler *et al.*, 1991), and discussed by Williams *et al.* (1996).

Differences in level and incidence of expression with developmental stage is no doubt a result of an interaction of factors including transgene degradation and replication, and the likelihood of some expression of chromosomally integrated copies with increasing developmental age.

The poor expression of the rat construct is in agreement with a number of previous studies. It has been reported by some researchers that mammalian-derived regulatory sequences and introns are frequently less efficiently transcribed in fish systems when compared to homologous sequences (e.g. Friedenreich & Schartl, 1990; Bearzotti *et al.*, 1992; Betancourt *et al.*, 1993). It appears, therefore, that the lower level of expression of the rat construct in both tilapia and trout may be due to an inherent species specificity of transcription regulatory factors. The observations that rainbow trout MTI and II promoter sequences have a much higher activity in fish cells than in other heterologous cells (Hong *et al.*, 1992) and in our studies that there is a higher level of expression of the ratβAlacZ than the carpβA*lacZ* construct in mouse cells (S. Morley, personal communication), are also in agreement with this view.

Conclusion

Taken together, the results presented here provide strong evidence that the carp β-actin regulatory sequences drive *lacZ* expression to levels several-fold higher than the equivalent rat sequences. Further work is currently under way in our laboratory to determine the regions conferring this species specificity, and also to discover the tissue location of the transient expression in the embryos.

Acknowledgements

We are most grateful to Dr Louise Lavender for useful discussion. We would also like to acknowledge financial support from the BBSRC and The Association of Commonwealth Universities (PhD studentship to Alam, M.S.)

References

Alam, M.S., Lavender, F.L., Iyengar, A., Rahman, M.A., Ayad, H.H., Lathe, R., Morley, S.D. & Maclean, N. Comparison of the activity of carp and rat B-actin gene regulatory sequences in tilapia and rainbow trout embryos (submitted for publication).

Bearzotti, M., Perrot, E., Michard-Vanhee, C., Jolivet, G., Attal, J., Theron, M.-C., Puissant, C., Dreano, M., Kopchik, J.J., Powell, R., Gannon, F., Houdebine, L.-M. & Chourrout, D. (1992). Gene expression following transfection of fish cells. *Journal of Biotechnology*, **26**, 315–25.

Betancourt, O.H., Attal, J., Theron, M.C., Puissant, C. & Houdebine, L.M. (1993). Efficiency of introns from various origins in fish cells. *Molecular Marine Biology, Biotechnology*, **2**, 181–8.

Braell, W.A. (1991). β-Galactosidase assay using the TKO100 minifluorometer. *Hoefer Scientific Inc. Technical Bulletin No. 29*, 4p.

Chong, S.S.C. & Vielkind, J.R. (1989). Expression and fate of CAT reporter gene microinjected into fertilized medaka (*Oryzias latipes*) eggs in the form of plasmid DNA, recombinant phage particles and its DNA. *Theoretical Applied Genetics*, **78**, 369–80.

Friedenreich, H. & Schartl, M. (1990). Transient expression directed by homologous and heterologous promoter and enhancer sequences in fish cells. *Nucleic Acids Research*, **18**, 3299–305.

Gong, G. & Hew, C.L. (1991). Functional analysis and temporal expression of promoter regions from fish antifreeze protein genes in transgenic medaka embryos. *Molecular Marine Biology and Biotechnology*, **1**, 64–72.

Hong, Y., Winkler, C., Brem, G. & Schartl, M. (1992). Development of a heavy metal-inducible fish specific expression vector for gene transfer *in vitro* and *in vivo*. *Aquaculture*, **111**.

Iyengar, A., Müller, F. & Maclean, N. (1996). Regulation and expression of transgenes in fish. *Transgenic Research*. In press.

Liu, Z.J., Moav, B., Faras, A.J., Guise, K.S., Kapuscinski, A.R. & Hackett, P.B. (1990*a*). Functional analysis of elements affecting expression of the β-actin gene of carp. *Molecular Cellular Biology*, **10**, 3432–40.

Liu, Z.J., Zhu, Z., Roberg, K., Faras, A.J., Guise, K.S., Kapuscinski, A.R. & Hackett, P.B. (1990*b*). Isolation and characterization of β-actin gene of carp (*Cyprinus carpio*). DNA Sequence. *Journal of DNA sequencing and Mapping*, **1**, 125–36.

Liu, Z.J., Moav, B., Faras, A.J., Guise, K.S., Kapuscinski, A.R. & Hackett, P.B. (1991). Importance of the CArG box in regulation of β-actin encoding genes. *Gene*, **108**, 211–18.

Moav, B., Liu, Z., Caldovic, L.D., Gross, M.L., Faras, A.J. & Hackett, P.B. (1993). Regulation of expression of transgenes in developing fish. *Transgenic Research*, **2**, 153–61.

Nudel, U., Zakut, R., Shani, M., Neuman, S., Levy, Z. & Yaffe, D. (1983). The nucleotide-sequence of the rat cytoplasmic β-actin gene. *Nucleic Acids Research*, **11**, 1759–71.

Penman, D.J., Iyengar, A., Beeching, A.J., Rahman, A., Sulaiman, Z. & Maclean, N. (1991). Patterns of transgene inheritance in rainbow trout (*Oncorhynchus mykiss*). *Molecular Reproductive Development*, **30**, 201–6.

Rahman, M.A. & Maclean, N. (1992). Production of transgenic tilapia (*Oreochromis niloticus*) by one-cell stage microinjection. *Aquaculture*, **105**, 219–32.

Stuart, G.W., McMurray, J.V. & Westerfield, M. (1988). Replication, integration and stable germline transmission of foreign sequences injected into early zebrafish embryos. *Development*, **103**, 403–12.

Stuart, G.W., Vielkind, J.R., McMurray, J.V. & Westerfield, M. (1990). Stable lines of transgenic zebrafish exhibit reproducible patterns of transgene expression. *Development*, **109**, 577–84.

Williams, D.W., Müller, F., Lavender, F.L., Orban, L. & Maclean, N. (1996). High transgene activity in the yolk syncitial layer affects quantitative transient expression assays in zebrafish embryos. *Transgenic Research*. (In Press)

Winkler, C., Vielkind, J.R. & Schartl, M. (1991). Transient expression of foreign DNA during embryonic and larval development of the medaka fish (*Oryzias latipes*). *Molecular and General Genetics*, **226**, 129–40.

Winkler, C., Hong, Y., Witbrodt, J. & Schartl, M. (1992). Analysis of heterologous and homologous promoters and enhancers *in vitro* and *in vivo* by gene transfer in Japanese medaka (*Oryzias latipes*) and *Xiphophorous*. *Molecular Marine Biology and Biotechnology*, **1**, 326–37.

F. MÜLLER, L. GAUVRY, D.W. WILLIAMS, J. KOBOLÁK, N. MACLEAN, L. ORBAN and G. GOLDSPINK

The use of transient *lacZ* expression in fish embryos for comparative analysis of cloned regulatory elements

Transgenic fish in applied and basic research

The introduction of foreign genes into fish embryos and the subsequent production of transgenic fish, have by now resulted in the accumulation of a substantial amount of data worldwide and there exists a large literature on the subject (for reviews see Maclean, Penman & Zhu, 1987; Fletcher & Davies, 1991; Hackett, 1993). The two major objectives of these studies are the production of valuable broodstock benefiting from the acquisition of desirable characteristics as well as the study of vertebrate gene regulation and the genetic basis of development.

The production of transgenic fish with the desire to found genetically superior broodstock for food production has been an aim since Zhu and coworkers introduced a growth hormone gene into goldfish (1985). Attempts to increase the growth rate of farmfish species by introducing growth hormone genes have also been made by many research laboratories, some with success (Du *et al.*, 1992; Chen *et al.*, 1993; Devlin *et al.*, 1994). Other goals include increased freeze tolerance (Fletcher, Davies & Hew, 1992) or improved disease resistance (Leong, 1994) using transgenic technology.

Fish are also becoming a popular model for the study of development and gene regulation. One approach to the study of gene regulation is the isolation and characterization of regulatory sequences and subsequent testing of their functionality *in vivo* or *in vitro*. The number of genes isolated from fish grows steadily. In a survey carried out by Maclean and Rahman (1994), the isolation of over 70 cDNA and genomic DNA sequences from a wide range of fish species by the end of 1992 is listed. This number has increased substantially in the past two years, owing to the recent interest in medaka (*Oryzias latipes*) and mainly zebrafish (*Danio rerio*) as a vertebrate model for developmental genetics (Rossant & Hopkins, 1992), and has led to the identification of many genes playing important roles in development. The reason why zebrafish, this tiny tropical species, has become the model species is that it allows both the application and exploitation of experimental embryology and genetic analysis (Driever *et al.*, 1994). Transgenic zebrafish are produced in many laboratories (e.g. Stuart, McMurray & Westerfield, 1988; Lin, Yang & Hopkins, 1994*a*); however, the efficiency of zygote microinjection in creating transgenic lines is still much lower than that regularly found in mice.

Fish offer many advantages over mammalian systems such as high fecundity, external fertilization and development, in addition to the wide range of available and relatively easy manipulation techniques such as the production of gynogenetic and androgenetic lines or triploid sterile individuals (for review see Horváth & Orbán, 1995). Advances have also been made in gene transfer technologies and novel methods show promise to provide alternatives to the often tedious microinjection of fish eggs. These techniques include electroporation of eggs (Inoue *et al.*, 1990; Powers *et al.*, 1992; Müller *et al.*, 1993) or sperm (Müller *et al.*, 1992; Sin *et al.*, 1993), liposome-mediated gene transfer (Szelei *et al.*, 1994) or tungsten bombardment (Zelenin *et al.*, 1991). Efforts are being made to enhance the rare and mostly late integration of transgene in the host genome using purified integrase protein (Ivics, Izsvák & Hackett, 1993) or pseudotyped retrovirus vector (Lin *et al.*, 1994*b*). Many investigators have looked into the processes involved in transgene expression and regulation in fish (e.g. Chong & Vielkind, 1989; Gong, Hew & Vielkind, 1991; Moav *et al.*, 1993; Westerfield *et al.*, 1992).

Whether transgenic fish are produced for agricultural biotechnology or for studying gene regulation in development, it becomes important to maintain controlled gene expression. This requires characterizing and testing the functionality of the regulatory elements driving the gene of interest. Two distinct systems are used to investigate the functionality of gene constructs: cell culture and whole animal systems.

The advantage of introducing a reporter gene construct into an intact organism, e.g. a fish embryo, is that the gene expression can be monitored spatially and temporally during development and differentiation. Therefore, significantly more information may be obtained than in a cell culture where a single cell type response is measured which cannot mimic the conditions of the multicellular organism. Although the transformation of a cell culture also provides advantages (fast reproducible and quantitative results), it cannot substitute for the *in vivo* analysis.

Reporter genes isolated from prokaryotic sources have been used extensively in fish, e.g. the *E.coli* derived chloramphenicol acetyl transferase (CAT, Stuart *et al.*, 1990; Tewari *et al.*, 1992), neomycin phosphotransferase (neo, Guise, Hackett & Faras, 1992) and beta-galactosidase (*lacZ*, Inoue *et al.*, 1990; Bayer & Campos-Ortega, 1992) genes. Recent favourites are the firefly luciferase (Sekkali *et al.*, 1994) and the green fluorescent protein genes (Mihalik, personal communication). Perhaps the most versatile of the reporter genes, which has been used in many different kinds of expression assays in transgenic fish, is the *lacZ* gene. The activity of the product of the gene can be detected using histochemical staining of fixed whole mount embryos or tissues based on the hydrolysis of 5-bromo-4-chloro-3-indoxyl-β-D-galactoside (X-gal) into galactose and soluble indoxyl molecules which are, in turn, oxidized to insoluble indigo, thus rendering expressing cells blue (MacGregor *et al.*, 1991). This protocol allows *in situ* detection in fixed embryos and is frequently used in studies on promoter/enhancer activity (e.g. Bayer & Campos-Ortega, 1992; Rinder *et al.*, 1992; Reinhardt *et al.*, 1994). Other *in situ* β-gal detection methods such as the application of the commercially available antibody (Sigma) to β-gal for immunochemical localisation are also possible in fixed zebrafish embryos (Williams *et al.*, in preparation).

More recently, a series of lipophilic fluorogenic substrates (fluorescein-di-β-D-galactopyranoside or FDG) are available which are readily taken up by cells and their fluorescence can be detected in the living embryos *in vivo*. This method has already been shown to work with zebrafish embryos using a low power fluorescence microscope (Lin *et al.*, 1994*a*). A commercially available FDG substrate ('ImaGene'–Molecular Probes) has also been utilized by Westerfield *et al.* (1992), where a low-light video camera was used to visualize expression of *lacZ* in individual neurons of living zebrafish embryos.

Two major types of transgene expression can be distinguished in a fish system. The F0 generation of treated eggs may show strong expression from a gene construct without actually integrating the sequence into their genome, but maintaining the gene copies extra-chromosomally. This transient expression has been utilized in gene expression studies (for review see Vielkind, 1992); however, when the goal is the production of stable transgenic lines, the gene expression of the F1 generations maintained from gene copies integrated in the genome is of primary interest.

Gene expression in stable transgenic lines

It has been known for a long time that changes in the chromosomal location of a gene can inhibit or modulate its expression, i.e. cause a position effect (Lewis, 1950). Following random integration, it has frequently been observed that transgene expression varies depending on the chromosomal integration site (e.g. Sands *et al.*, 1993; Clark *et al.*, 1993 in transgenic mice; Stuart *et al.*, 1990 in transgenic fish). In addition to variable patterns of expression, position effects are also thought to be one of the factors responsible for the lack of correlation observed between transgene copy number and the levels of expression obtained. Such position effects are thought to be most strongly exerted on transgenes which do not contain strong promoters or enhancers (Allen *et al.*, 1988). Indeed, in a technique called 'enhancer trapping', such constructs are widely used to help identify hitherto unknown regulatory elements situated in proximity to the transgene in mouse (Skarnes, 1990).

It has also become clear recently that the eukaryotic genome is organized into topologically constrained loops, which are the basic regulatory units for the control of gene activity (e.g. Sippel *et al.*, 1992). Using the chicken lysozyme gene locus as an example, Sippel *et al.* (1992) observed that a number of regulatory elements within a domain of approximately 20 kb were involved in the activation of this gene. This regulatory unit referred to as a 'regulon', was found to contain at its boundaries stretches of DNA which are thought to attach to the chromosomal scaffold or to the nuclear matrix material at the base of the chromosomal loop and are referred to as matrix attachment regions (MARs) or scaffold attachment regions (SARs). These types of DNA regions have been found to insulate transgenes from position effects by a number of researchers (Chung, Whiteley & Felsenfeld, 1993; Kalos & Fournier, 1995). Upon coinjecting the MAR sequences from the chicken lysozyme locus (A element) with a highly position-dependent gene (whey acidic protein gene) and producing transgenic mice, McKnight *et al.* (1992) observed that a few of the transgenic lines produced exhibited position-independent expression of the gene. Hackett *et al.* (1994) have also reported enhanced levels of expression in zebrafish cell lines of a β-actin/CAT construct flanked by these sequences when compared to the same construct without matrix attachment region sequences. Finally, using transgenic zebrafish lines, Betancourt *et al.* (1993) have observed enhanced expression of viral driven CAT constructs using SV40 MAR sequences. Boundary regions of the *hsp*70 gene of *Drosophila* (specialized chromatin structure or 'scs elements'), which are also known to insulate transgenes from position effects, however, do not attach to the nuclear matrix (Kellum & Schedl, 1992).

It has also been found that the chicken lysozyme MAR sequences may insulate a strong enhancer from taking effect on the function of a promoter in transient conditions when the MAR sequences were placed in between the enhancer and the promoter regions. This was confirmed in transient transfection assays in macrophages by Stief and coworkers (1989).

Novel DNA often becomes *de novo* methylated and consequently often inactivated upon integration into an animal's genome. Nilsson & Lendahl (1993) have reported that a β-actin/*lacZ* construct, upon injection into mouse eggs, expressed transiently in *in vitro* cultured preimplantation embryo, but upon integration into five separate lines of transgenic mice became methylated and inactivated. The mechanisms of *de novo* methylation remain unclear, but it has been suggested that it is a host defence mechanism against foreign DNA expression upon failure of previous lines of defence such as degradation and excision (Doerfler, 1992). Another possible explanation for the genomic silencing of transgene is the formation of local heterochromatin which is initiated by the pairing of closely linked transgene copies as observed in *Drosophila* by Dorer & Henikoff (1994).

Transient expression

When a transgene is introduced into a fish embryo, the transient activity of the gene is mainly defined by its fate. The exogenous DNA introduced in either circular or linear form is most often not integrated into the genome but persists extrachromosomally, mainly in large molecular weight concatamers and often gets replicated (Chong & Vielkind, 1989). The expression from this DNA is called transient expression. This activity is usually high in the early developmental stages and may be retained for several weeks, until the DNA is finally degraded (Gong *et al.*, 1991). Although transient expression is not affected by position effects, the activity might be affected by further factors other than the regulator.

Unlike in stable lines of transgenics, copy number of the transgene is one of these factors in cells expressing extrachromosomal DNA. Although most experimenters attempt injection of identical copy numbers of DNA into every egg for their studies, variations in the amounts of injected DNA have been found to affect the subsequent levels of transient expression in fish eggs. Volckaert *et al.* (1994) have observed a good correlation between the levels of transient expression obtained and the copy number of the construct introduced. It is advisable to adopt a rigid DNA quantification protocol for such studies. In addition, if quantitative evaluation of promoter activities and comparative studies are undertaken, it is important for statistical comparisons that a high number of repetitions are carried out. Attempts are being made in our laboratories to eliminate or to control variability in DNA injection by adopting coinjection techniques (Williams *et al.*, in preparation).

The copy number of DNA introduced is mainly dependent on the method of gene delivery. When using microinjection of eggs, the volume injected may be controlled to some extent by adjusting the injection pressure of the microinjector device used; however, large egg to egg variations may still be observed. When using alternative technologies such as egg electroporation, there is no indication at all about the actual amount of DNA introduced. Transient expression data show that the levels of expression gained in electroporated embryos are usually smaller than in the case of microinjected embryos (Müller *et al.*, 1993), although the quantities of DNA used are orders of magnitude higher (mg as opposed to ng and pg quantities), however, this might be improved by optimizing the conditions of electroporation.

The other factor which affects transient expression seriously is the segregation of extrachromosomal DNA during cleavage of the blastomeres and their subsequent differentiation. It has been shown that novel and unintegrated DNA is unevenly distributed into the dividing cells, therefore its appearance and persistence in the different tissues will be varying (Westerfield *et al.*, 1992). As the tissue-specific regulatory sequences are not always active in all tissue types, the spatial expression pattern will be influenced by this phenomenon (Williams, in preparation). Again, using large numbers of embryos and statistical analysis may help in getting a more realistic picture defined by the regulator rather than DNA fate.

The study of transient expression in analyses of tissue specific promoters

The study of the spatial expression pattern of transgenes allows investigation of the tissue specificity of desired regulatory elements. Many such sequences have been studied in transiently transgenic fish and especially in the zebrafish and medaka species. In an early study, Ozato *et al.* (1986) injected the chicken-crystallin gene into medaka oocytes, and upon investigating the spatial patterns of expression in the embryos, observed expression in a number of different tissues with no apparent lens specificity.

More recently, however, Rinder *et al.* (1992) observed that the regulatory sequences from the zebrafish ependymin gene (the predominant glycoprotein in the cerebrospinal fluid of fish, thought to be involved in cell adhesion phenomena) when spliced to the *lacZ* gene, drove expression in an ependymin-specific manner in zebrafish embryos. Similarly, regulatory sequences from the rat *GAP* 43 gene (which codes for a major component of neuronal growth cones) driving the *lacZ* gene, were found to be specifically activated in the embryonic nervous system at the time of extensive neuronal differentiation in transgenic zebrafish embryos (Reinhardt *et al.*, 1994). This expression was also found to be down-regulated in the fish spinal cord, and found to increase in more rostral regions of the central nervous system, a pattern very similar to that of the endogenous fish *GAP* 43 gene.

Genes playing important roles in pattern formation during animal development such as the Hox genes have also been studied by Westerfield *et al.* (1992). These authors injected the mouse Hox-1.1 and human Hox-3.3 promoters spliced to a *lacZ* reporter gene and observed that they were activated in specific regions of the developing embryo showing conserved expression patterns comparable to those in mammals. The specificity of these sequences was, therefore, seen to be appropriately maintained. Ubiquitous promoters on the other hand, which are expected to be expressed throughout the embryo, are often found to exhibit non-uniform expression (e.g. Stuart *et al.*, 1990). This has been attributed to mosaicism of the transgene as a result of unequal distribution of the transgene copies (for review see Fletcher & Davies, 1991).

Experiments on studying muscle-specific transgene expression in fish embryos

Fish are ideal models for studying myogenesis and muscle development since most species develop rapidly and *ex utero*. Myotomes with their distinctive shape, can be easily identified by one day post-fertilization in zebrafish, and become striated by 36 hours (Kimmel & Warga, 1987; Felsenfeld, Curry & Kimmel, 1991). Muscle constitutes most of the mass of the trunk or tail of fish therefore the majority of muscle cell populations are accessible for observation in the transparent zebrafish embryo.

Although further studies have been published on fish somite and muscle development (Felsenfeld *et al.*, 1990; Halpern *et al.*, 1993), the number of publications on muscle genes and their regulation in fish is scarce. This is particularly so in comparison to other vertebrate systems such as mouse or chicken where the structure and expression of myosin genes in developing skeletal muscle have been well characterized (Weydert *et al.*, 1987; Kropp, Gulick & Robbins, 1987).

Myosin is a contractile protein which assembled into filaments and linked by crossbridges with actin filaments provides ATP-dependent muscular contraction. It is a major component of muscle cells and is encoded by members of a multigene family that are expressed in a developmentally and spatially regulated manner maintained mainly at the transcriptional level. The specific sequences that are responsible for the binding of MRFs have been characterized including the E box (Edmondson *et al.*, 1992) and other binding sites which have been found to be active in cardiac muscle genes such as the CArG box (Mohun *et al.*, 1989) and thyroid responsive elements (Subramaniam *et al.*, 1992).

Recently, a number of carp (*Cyprinus carpio*) myosin heavy chain gene isoforms have been isolated and their expression has been monitored in carp (Ennion *et al.*, 1995). The 5′ regulatory sequences of an adult isoform have also been isolated and characterized by splicing to a CAT reporter gene and subsequent transformation of mouse myoblast cells or direct microinjection into fish muscle (Gauvry *et al.*, 1995). It was found that an E-box like element is recognizable at the 5′ end of the regulator.

In the following paragraphs, experiments will be described which were undertaken in our laboratories aiming at the comparative analysis of this fish-derived tissue-specific regulator using transient expression in embryos of carp, African catfish (*Clarias gariepinus*) and zebrafish. The effect of enhancer and MAR sequences on the tissue specificity and level of expression have also been investigated and a novel and very simple method for facilitating comparative analysis of the regulators is described. A reporter construct containing the chicken lysozyme MAR sequences flanking the *lacZ* gene (provided by Prof. Bonifer, Heidelberg, FRG) was used as vector into which the carp-derived myosin heavy chain promoter (MyHC; 0.9 kb) has been inserted. With different restriction digestions, both the shorter fragment (4.9 kb) containing the promoter and reporter gene (MZ fragment) as well as the longer sequence (11.4 kb) containing the two MAR elements in inverted repeat form (3 kb each) could also be isolated (MMZM fragment). The rat-derived myosin light chain enhancer element (MyLC) was also used in these experiments. This 0.9 kb sequence was isolated by Donoghue *et al.* (1988) and was cut out from an expression vector for our studies (ME fragment).

A constitutive and general enhancer/promoter has also been used as control in comparison of tissue specificity of the regulators. The regulatory elements of the β-actin gene, an evolutionarily conserved, highly expressed sequence, which were isolated from the genome of common carp, were used. Extensive functional analyses on the putative control regions spliced to the CAT reporter gene have been carried out earlier after transfection into fish cells (Liu *et al.*, 1990). Moav *et al.* (1993) have obtained efficient expression of this regulator *in vivo* in zebrafish and goldfish. Recently, Alam, Popplewell & Maclean (1995) have shown its strong and general, but also position affected activity in stable transgenic lines of tilapia (*Oreochromis niloticus*).

The 5.9 kb fragment of the 5′ regulatory sequences of the β-actin promoter containing the untranslated first exon, the first intron (found to possess an enhancer) and part of the second exon in addition to the proximal promoter spliced to the *lacZ* reporter gene was used in our transient expression assays (8.9 kb). In coinjection experiments the isolated regulator without the *lacZ* gene was used (3.2 kb CBA fragment).

Transient expression studies in three species showing similar tissue specificity

Embryos microinjected with the MZ fragment were examined for *lacZ* activity in their cells by whole mount X-gal staining. Numbers of expressing cells were usually very low, due to transgene mosaicism and probably the weakness of the promoter. Positive cells were identified and counted. A simple sketch of an embryo showing all somites and major body areas was used to record the numbers and relative spatial position of positive cells and the data from all batches of treated individuals were accumulated onto a single drawing, and produced as a printout. The total number of expressing cells (or clusters of cells) per positive embryo was used as indication of expression levels. The tissue specificity of expression was calculated as the mean ratio of the number of expressing muscle cells (or clusters) to the total number of expressing cells in a batch of embryos. The incidence of expressing individuals in relation to the total number of individuals analysed was expressed as a percentage. The frequency distribution of the number of expressing cells per embryo was calculated and demonstrated on a bar chart diagram.

Expression was assayed in carp, zebrafish and catfish embryos and larvae at different stages of development including 80% epiboly, tail bud stages and at 1, 3, 5, 10 days. Analysis of individuals of carp was not possible from 10 days onwards owing to the thick tissues which covered the stained cells and prevented *in situ* assessment of *lacZ* activity. The first visible X-gal staining signal was detectable at 1 day of age in all three species. The expression patterns were mosaic, few cells were shown to be expressing β-galactosidase at all stages. After four series of injections, 275 one-day-old carp embryos (prehatch stage), 215 three-day-old larvae (free swimming but yolk sac is still found), and 122 five-day-old larvae (feeding larvae stage) were assayed. Development of the embryos was rapid due to the relatively high incubation temperature (28 °C). No abnormalities or developmental differences were observed when compared with mock injected or non-injected controls incubated at 22 °C (data not shown). The number of expressing cells in each treatment was low, with only a small proportion of fibres showing β-galactosidase activity, partly due to mosaicism at the DNA level.

In the microinjected fish, expression was mostly found in the skeletal muscle fibres. Only a few other types of cells (e.g. the extraembryonic tissues or a group of cells in the head region) were rarely found to express *lacZ* (83.9% specificity). The distribution of expressing muscle cells appeared random, with no preferential areas. No expression was found in other types of muscle, i.e. in the heart or visceral mesoderm. By the third day of development, all non-specific expression disappeared and only muscle cells exhibited expression of the MyHC/lacZ construct. The typical expression pattern found in carp at 1 day of development is shown in Fig. 1. There was considerable variation in the frequency of expression among the individuals (21–77%), but the mean data of the three developmental stages showed no significant difference (41.5%, 44.8% and 43.5%) at 1, 3 and 5 days of development, respectively.

The number of β-galactosidase expressing cells (levels of expression) increased slightly as development progressed (from mean 3.8 to 6.1 cells per embryo, at 1 and 5 days, respectively). The mean frequency of expressing zebrafish and catfish individuals was 23%, and 32% at one day age, respectively. The mean level of expression in the zebrafish embryos was very low 1.6 ± 0.9, whereas the specificity of expression was 74.4%. A newly hatched zebrafish embryo showing expression of β-galactosidase in its myotomes is shown in Fig. 2. In catfish, the average number of expressing cells was found to reach an average 3.9/embryo and the specificity of expression was similar to zebrafish giving a mean of 82.6%. The tissue-specific appearance of expression was caused by the promoter itself rather than transgene mosaicism which was shown in a comparative experiment in catfish where a *lacZ* gene driven by a general, constitutive promoter was also used (Table 1.). A batch of catfish embryos were injected with a carp β-actin/*lacZ* fragment to allow comparison of the expression pattern with that of

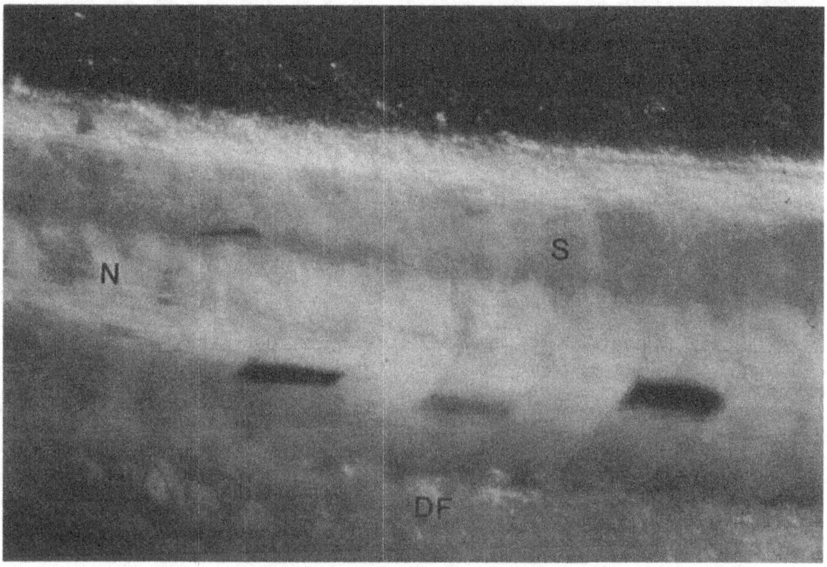

Fig. 1. Expression of MyHC/lacZ in myotomes of X-gal stained 3-day-old carp larva. Abbreviations: S: somites, N: notochord, DF: dorsal fin, VF: ventral fin.

MyHC/*lacZ*. The number of expressing muscle cells as a percentage of the total number of expressing cells in positive individuals (4.4% in case of the β-actin regulator, while 90.3% in the control batch injected with the myosin promoter construct) demonstrated the high level of tissue specificity of the cMyHC promoter (Table 1) as compared to the general promoter.

The fact that the carp MyHC promoter caused tissue-specific expression indicates that its sequence contains the elements required for targeting the expression into muscle cells. According to Gauvry *et al.* (1995) the regulatory sequence contains not only the CAAT and TATA boxes, but an E box (CANNTG) at −896 bp which could be responsible for the muscle-specific expression shown by this promoter.

The reason for the low level of expression may be explained by several possibilities. The endogenous gene from which the promoter was isolated was found to be an isoform expressed mainly in adult carp (Ennion *et al.*, 1995) and may not be fully active in embryonic stages. Another possibility is that further promoter elements and enhancers are lacking from the 0.9 kb promoter fragment as it has

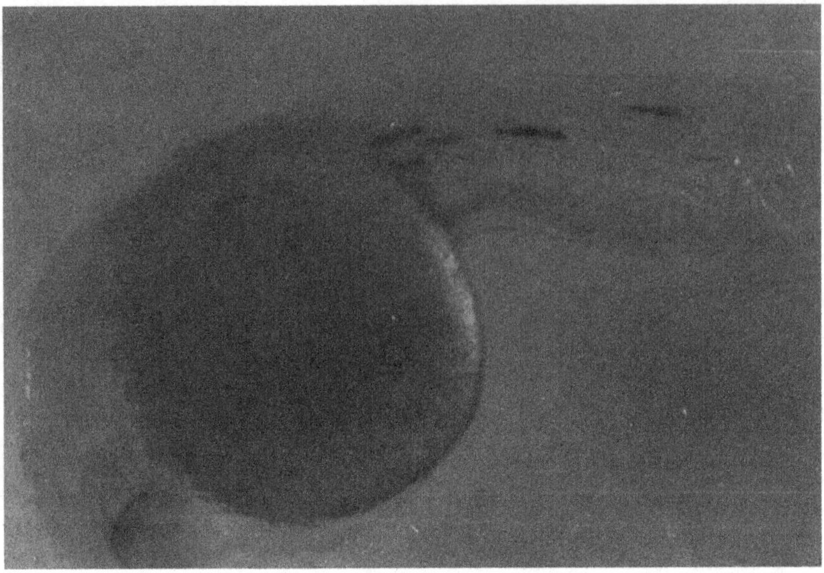

Fig. 2. Expression of MyHC/lacZ in cells of newly hatched 1-day-old zebrafish embryo.

Table 1. *Comparison and distribution of expression of lacZ driven by a carp β-actin regulator and a carp myosin heavy chain promoter in African catfish larvae.*

Regulator	Frequency of positives	%	Muscle cell/ +individual	All positive cells/ +individual	Ratio (%)
carp beta-actin	13/23	56.5	0.92	20.76	4.4
carp MyHC	12/30	40	2.33	2.58	90.3

Abbreviations: muscle/+ individual: β-galactosidase expressing skeletal muscle cells per positive individual; total+/+individual: total number of β-galactosidase expressing cells per positive individual; muscle spec. (%): ratio of positive muscle cells per total number of positive cells in percentage.

been shown for the rat embryonic MyHC gene (Bouvagnet *et al.*, 1987). This possibility is partly justified in the experiments described in the next chapters below.

Although the endogenous isoform of cMyHC was also found to be temperature dependent (Ennion *et al.*, 1995), however, our experiments did not show striking differences between embryos incubated at 17 °C or 28 °C (data not shown). This could be the result of using a truncated form of the regulator sequence in these experiments.

Enhancer sequences can take specific effect on transient expression when coinjected as separate fragments

To assess whether a rat-derived myosin light chain enhancer element may confer elevated tissue-specific expression of the transgene during embryogenesis, the MyLC enhancer fragment (ME) was coinjected with the MyoHC/*lacZ* fragment (MZ) into zebrafish early embryos. Hatched embryos were fixed and analysed at 1 day of age. Among the eight batches of MyHC/*lacZ* injected zebrafish with a total number of 207 individuals analysed, 26 were found to express β-galactosidase, whereas, out of 289 coinjected embryos (MZ+ME), 133 showed positive signal. The level of expression was higher in the co-injected batches (mean 6.0 ± 3.0 as compared with 1.8 ± 0.6) than in those injected with the promoter construct alone (significant difference at 95% probability, u=42.4 at z=1.960). The muscle specificity of expression was somewhat decreased, but still retained at high level (75 ± 12%). When this coinjection experiment was performed in catfish, the results were similar to those obtained with zebrafish (data not shown).

The 3.2 kb fragment containing a constitutive carp β-actin regulator (CBA) was coinjected together with the MyHC/*lacZ* fragment (MZ) in four batches to test and compare the tissue specificity of the enhancer effect. A total number of 190 1-day-old zebrafish embryos were analysed and 24% were found to express the *lacZ* gene. The level of expression was increased (significant difference at 95% probability, u=72.6 at z=1.960); however, the enhancement was found mainly aspecific, not restricted to muscle cells. The ratio of cells expression in skeletal muscle was much lower (14%) than in the case of the MZ+ME coinjection experiments. The differences in the expression levels and patterns obtained from these coinjection experiments are demonstrated in Fig. 3.

The addition of a heterologous muscle-specific enhancer or a constitutive carp regulator by coinjection significantly improved the levels of *lacZ* expression measured in two fish species, although the enhancers

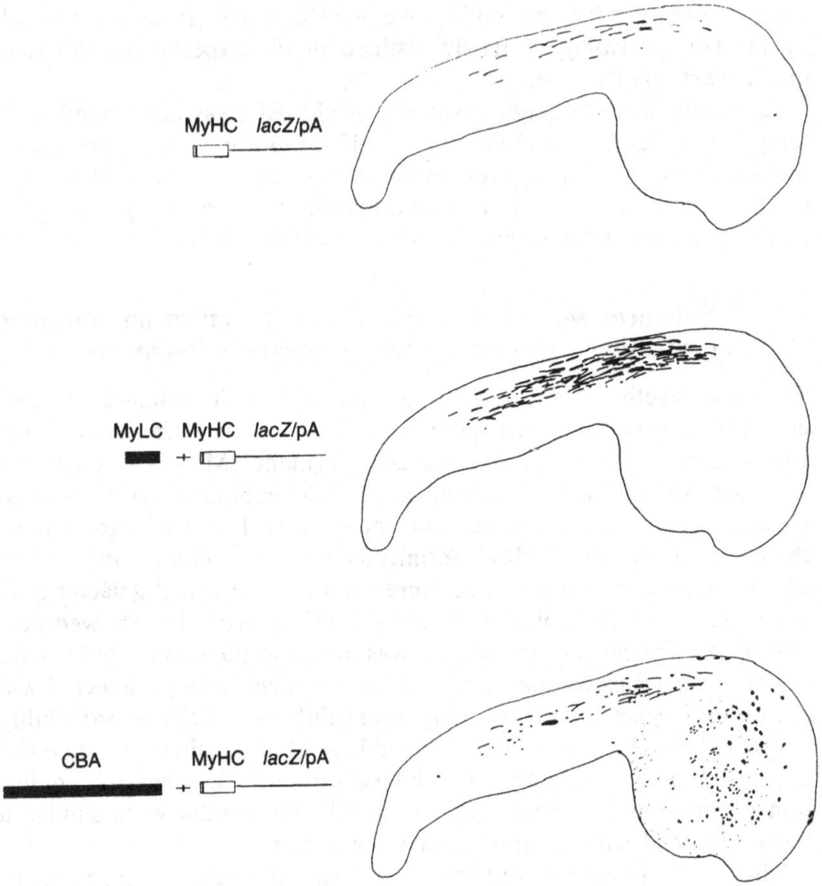

Fig. 3. *LacZ* gene expression maps of 1-day-old zebrafish embryos obtained from coinjection experiments (data collected from 207, 289 and 190 treated embryos, respectively). Abbreviations: MyHC: carp myosin heavy chain promoter, lacZ/pA: *E.coli lacZ* gene and SV40 poly(A) signal; MyLC: rat-derived myosin light chain enhancer fragment; CBA: carp β-actin regulatory region fragment.

were not covalently linked to the expression vector at the time of microinjection.

This effect of the enhancer may be explained in two ways. It is possible that the CBA enhancer fragments became involved in concatemer formation with the MZ fragment. In both *Xenopus* and fish, concatemers formed by circular injected DNA have been found to be largely of head-to-tail arrangements, whereas those formed by linear injected DNA (as in this case) have been found to be of random arrangements (e.g. Marini, Etkin & Benbow, 1988; Endean & Smithies, 1989; Vielkind, 1992). Marini *et al.* (1988) suggested that linear molecules rapidly ligated into long random concatemers in *Xenopus* due to the presence of large amounts of stored ligases in the egg. Although the presence of a large number of DNA 'ends' is known to stimulate the production of ligases (Bishop & Smith, 1989), concatemerization of linear molecules in *Xenopus* and fish has often been found to be very rapid (within 5 minutes of injection in medaka, Chong & Vielkind, 1989). This concatemer formation takes place even between fragments containing heterologous ends (Erdélyi, F., personal communication) and the reaction between the fragments used in this experiment could take place similarly.

The other possibility could be that the rat and carp enhancers acted *in trans*, i.e. not covalently linked to the MZ fragment. It has already been shown *in vitro* that enhancer fragments may produce an effect on a promoter located on a separate molecule *in trans*, supporting the looping model of enhancer action (Müller, Sogo & Schaffner, 1989). It was also found, in the fruitfly (*Drosophila melanogaster*), that a specific locus on one chromosome could influence gene expression in the homologous locus of the paired chromosome in the phenomenon called transvection, which may be explained by a similar *trans* effect (for review see Müller & Schaffner, 1990). Experiments are in progress to study the mechanism for enhancer activity in the fish embryos produced in these studies.

Although the coinjection of the rat enhancer increased the number of cells expressing *lacZ* by a factor of 3–7, the total number of positive cells per embryo still remained low. The reasons for this could be the following: (i) the enhancer is heterologous; (ii) further elements are lacking from the enhancer itself; (iii) the enhancer could not perform its effect with full capacity in the conditions of coinjection; (iv) the promoter is an adult myosin isoform, which might exhibit limited activity during the stages investigated; and (v) mosaicism at the DNA level as previously mentioned.

Influencing transgene expression by MAR elements

In these experiments the effect of the chicken lysozyme MAR sequences was assayed on the expression levels and patterns of the MyHC/*lacZ* construct. The fragment containing the two boundary sequences was injected into carp and zebrafish embryos and data were collected and assayed as mentioned above. These sequences had a surprising and marked effect on the expression levels and patterns. It was found that the tissue specificity of expression was somewhat lost, fewer numbers of muscle cells were found to be expressing β-gal, whereas a general enhancement in expression in other cell types was found, indicating that the MAR sequences might have an aspecific enhancer function as well. The ratio and frequency distribution of expressing cells per carp embryo are shown in Fig. 4.

The accumulated data collected from similarly treated zebrafish embryos are shown in Fig. 5. In this Figure, a third experimental result is also demonstrated where the ME fragment was coinjected with the MMZM fragment. It reveals that the rat-derived MLC enhancer could still take effect on the MyHC promoter activity by enhancing expression levels mainly in muscle cells, although a boundary sequence, the chicken lysozyme MAR element was flanking the reporter construct which should have insulated the coinjected regulator. This result contradicts that of Stief *et al.* (1989), where the enhancer used was not able to act through the MAR sequences when covalently linked on the reporter fragment in transient transfection experiment. The reason why the enhancer may have worked in our experiment could be the *trans* effect which was unlikely with the constructs of Stief and coworkers. Also, in their experiment cat reporter gene, expression was analysed using homogenates of transfected cell culture which approach could not reveal tissue specific differences.

Conclusions

It has been successfully demonstrated that a carp-derived MyHC promoter directs the expression of *lacZ* reporter gene in a skeletal muscle-specific manner not only in carp but in two other fish species in transient conditions. A considerable number of publications have appeared in

---

Fig. 4. Frequency distribution of *lacZ* gene expression in 1-day-old carp embryos. Numbers in parentheses represent the number of expressing individuals and the number of embryos analysed. Open bars represent the total number of expressing cells per embryo whereas the solid bars represent the number of expressing muscle cells per embryo. Abbreviations: MZ: embryos injected with cMyHC/lacZ fragment only; MMZM: MyHC/lacZ fragment flanked by the chicken lysozyme MAR sequences.

MZ (85/184)

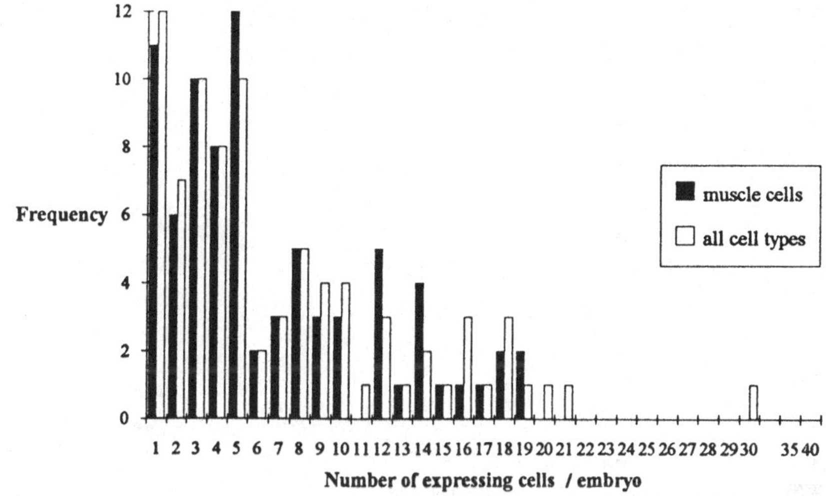

MMZM (60/132)

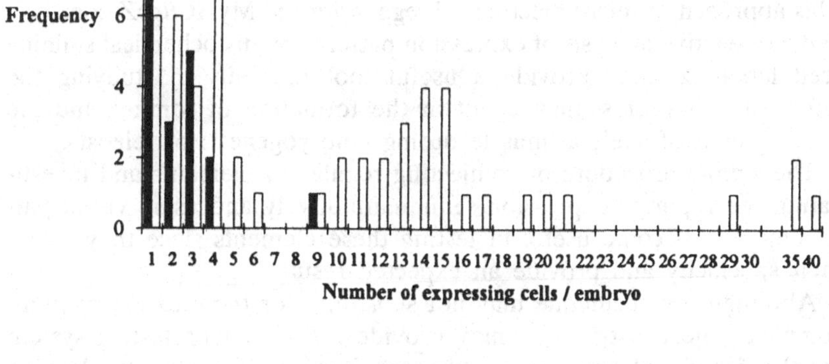

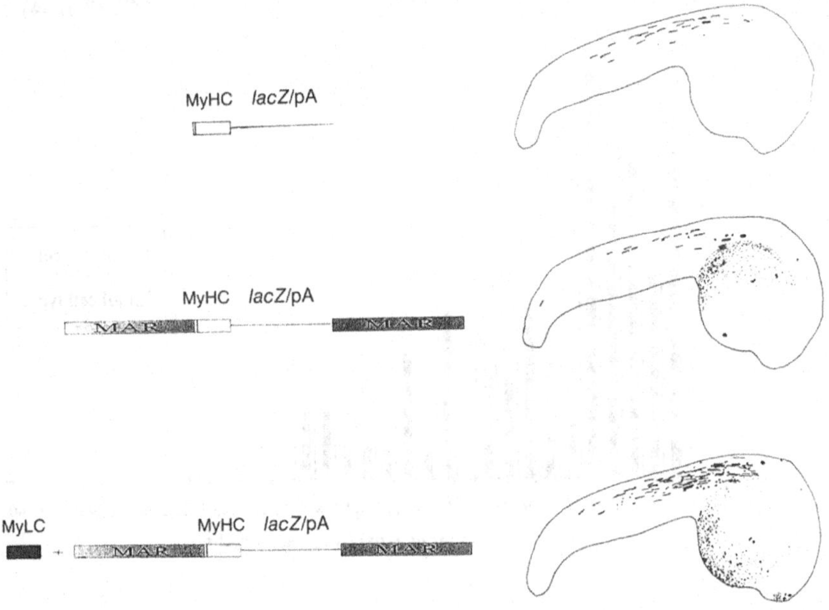

Fig. 5. *LacZ* expression map of 1 day old zebrafish embryos injected with different DNA fragments. Abbreviations: MyHC: carp myosin heavy chain promoter; lacZ/pA: *E.coli lacZ* gene and SV40 poly(A) signal; MyLC: rat-derived myosin light chain enhancer fragment; MAR: chicken lysozyme MAR sequences.

the literature concerning myogenesis in mice and chicken, and the structure and function of the genes involved including studies in transgenic systems (Petropoulos *et al.*, 1989; Rindt *et al.*, 1993). This is, however, the first report of similar experiments in transgenic fish. This approach of microinjecting fish eggs with the MyHC/*lacZ* construct and consecutive analysis of expression patterns by histochemical staining (cell labelling) may provide a useful tool not only in studying the effect of enhancer sequence but in the formation of somites and the development of skeletal muscle during embryogenesis of teleosts.

The simple procedure of coinjecting regulatory elements and investigating the reporter expression semiquantitatively and using visual patterning may become useful in testing these elements since they retain their specificity and provide an expected result.

Although this technique may not substitute for the creation of more complex gene constructs, it may provide a fast *in vivo* testing system for the functional analysis of enhancer/silencer elements. By cleaving these sequences into shorter regions and subsequent coinjection of these, more detailed information may be provided on the specific regions involved in tissue-specific enhancer activity (i.e. determination of transcription factor binding regions). By calculating the number of expressing cells, rate of tissue specificity may be assessed and fine differences may be revealed, which may not be possible by the quantitative gene expression analyses in tissue homogenates of embryos or tissue culture cells.

Acknowledgements

We are grateful for the help of Louise Lavender, József Burin and Sándor Egedi. This work is supported by a grants to G.G. from NFRC, AFRC and the Wellcome Trust, and by a grant to F.M. from the Hungarian Academy of Sciences (OTKA No. F013027).

References

Alam, M.S., Popplewell, A. & Maclean, N. (1996). Germ line transmission and expression of a *lacZ* containing transgene in tilapia. (*Oreochromis niloticus*). *Transgenic Research*, in press.

Allen, N.D., Cran, D.G., Barton, S., Hettle, S., Reik, W. & Surani, M.A. (1988). Transgenes as probes for active chromosomal domains in mouse development. *Nature*, **333**, 852–5.

Bayer, T.A. & Campos-Ortega, J.A. (1992). A transgene containing *lacZ* is expressed in primary sensory neurons in zebrafish. *Development*, **115**, 421–6.

Betancourt, O.H., Attal, J., Théron, M.C., Puissant, C. & Houdebine, L.M. (1993). Efficiency of introns from various origins in fish cells. *Molecular Marine Biology and Biotechnology*, **2**, 181–8.

Bishop, J.O. & Smith, P. (1989). Mechanism of chromosomal integration of microinjected DNA. *Molecular and Biological Medicine*, **6**, 283–98.

Bouvagnet, P.F., Strehler, E.E., White, G.E., Strahler-Page, M.A., Nadal-Ginard, B. & Mahdavi, V. (1987). Multiple positive and negative 5' regulatory elements control the cell-type-specific expression of the embryonic skeletal myosin heavy-chain gene. *Molecular and Cellular Biology*, **7**, 4377–89.

Chen, T.T., Kight, K.K., Lin, C.M., Powers, D.A., Hayat, M., Chatakondi, N., Ramboux, A.C., Duncan, P.L. & Dunham, R.A. (1993). Expression and inheritance of RSVLTR-rtGH1 complementary DNA in transgenic common carp, *Cyprinus carpio*. *Molecular Marine Biology and Biotechnology*, **2**, 88–95.

Chong, S.S.C. & Vielkind, J.E. (1989). Expression and fate of CAT reporter gene microinjected into fertilised medaka (*Oryzias latipes*) eggs in the form of plasmid DNA, recombinant phage particles and its DNA. *Theoretical and Applied Genetics*, **78**, 369–80.

Chung, J.H., Whiteley, M. & Felsenfeld, G. (1993). A 5′ element of the chicken β-globin domain serves as an insulator in human erythroid cells and protects against position effect in *Drosophila. Cell*, **74**, 505–14.

Clark, A.J., Archibald, A.L., McClenagham, M., Simons, J.P., Wallace, R. & Whitelaw, C.B.A. (1993). Enhancing the efficiency of transgene expression. *Philosophical Transactions of the Royal Society of London B*, **339**, 225–32.

Devlin, R.H., Yesaki, T.Y., Biagi, C.A., Donaldson, E.M., Swanson, P. & Chan, W. (1994). Extraordinary salmon growth. *Nature*, **371**, 209–10.

Doerfler, W. (1992). DNA methylation: eukaryotic defense against the transcription of foreign genes. *Microbial Pathogenesis*, **12**, 1–8.

Donoghue, M., Ernst, H., Wentworth, B., Nadal-Ginard, B. & Rosenthal, N. (1988). A muscle-specific enhancer is located at the 3′ end of the myosin light-chain 1/3 gene locus. *Genes Development*, **2**, 1779–90.

Dorer, D.R. & Henikoff, S. (1994). Expansions of transgene repeats cause heterochromatin formation and gene silencing in *Drosophila. Cell*, **77**, 993–1002.

Driever, W., Stemple, D., Schier, A. & Solnica-Krezel, L. (1994). Zebrafish: genetic tools for studying vertebrate development. *Trends in Genetics*, **10**, 152–9.

Du, S.J., Gong, Z., Fletcher, G.L., Shears, M.A., King, M.J., Idler, D.R. & Hew, C.L. (1992). Growth enhancement in transgenic Atlantic salmon by the use of an 'all-fish' chimeric growth hormone gene construct. *Biotechnology*, **10**, 176–81.

Edmondson, D.G. & Olson, E.N. (1989). A gene with homology to the myc similarity region of MyoD1 is expressed during myogenesis and is sufficient to activate the muscsle differentiation program. *Genes Development*, **3**, 628–40.

Edmondson, D.G., Cheng, T.C., Cserjesi, P., Chakraborty, T. & Olson, E.N. (1992). Analysis of the myogenin promoter reveals an indirect pathway for positive autoregulation mediated by the muscle specific enhancer factor MEF2. *Molecular and Cellular Biology*, **12**, 3665–7.

Ennion, S., Gauvry, L., Butterworth, P. & Goldspink, G. (1995). Small diameter white muscle fibres associated with growth hyperplasia in the carp (*Cyprinus carpio*) express a distinct myosin heavy chain gene. Submitted to *Journal of Biological Chemistry*.

Endean, D. & Smithies, O. (1989). Replication of plasmid DNA in fertilized Xenopus eggs is sensitive to both the topology and size of the injected template. *Chromosoma*, **97**, 307–14.

Felsenfeld, A., Walker, C., Westerfield, M., Kimmel, C. & Streisinger, G. (1990). The *fub-1* mutation blocks initial myofibril formation in zebrafish muscle pioneer cells. *Development*, **108**, 443–59.

Felsenfeld, A., Curry, M. & Kimmel, C.B. (1991). Mutations affecting skeletal muscle myofibril structure in the zebrafish. *Developmental Biology*, **48**, 23–30.

Fletcher, G.L. & Davies, P.L. (1991). Transgenic fish for aquaculture. In *Genetic Engineering*, vol. 13, ed. J.K. Setlow, pp. 331–370. New York: Plenum Press.

Fletcher, G.L., Davies, P.L. & Hew, C.L. (1992). Genetic engineering of freeze-resistant Atlantic salmon. In *Transgenic Fish*, ed. C.L. Hew & G.L. Fletcher, pp. 190–208. World Scientific.

Gauvry, L., Ennion, S., Hansen, E., Butterworth, P. & Goldspink, G. (1995). Cloning and characterisation of a carp myosin heavy chain gene promoter region. *Journal of Biological Chemistry* (submitted).

Gong, Z., Hew, C.L. & Vielkind, J.R. (1991). Functional analysis and temporal expression of promoter regions from fish antifreeze protein genes in transgenic Japanese medaka embryos. *Molecular Marine Biology and Biotechnology*, **1**, 64–72.

Guise, K.S., Hackett, P.B. & Faras, A.J. (1992). Transfer of genes encoding neomycin resistance, chloramphenicol acetyl transferase and growth hormone into goldfish and northern pike. In *Transgenic Fish*, ed. C.L. Hew & G.L. Fletcher, pp. 142–163, World Scientific.

Hackett, P.B. (1993). The molecular biology of transgenic fish. In *Biochemistry and Molecular Biology of Fishes*, vol. 2, ed. P.W. Hochachka & T.M. Mommsen, pp. 207–240. Elsevier.

Hackett, P.B., Caldovic, L., Izsvák, Z., Ivics, Z., Fahrenkrug, S., Kaufman, C., Martinez, G. & Essner, J.E. (1994). Vectors with position-independent expression and enhanced transfer for production of transgenic fish. Abstract. p. 73 in *Program and Abstracts*, IMBC, August 7–12, 1994, Tromsö, Norway.

Halpern, M.E., Ho, R.K., Walker, C. & Kimmel, C. (1993). Induction of muscle pioneers and floor plate is distinguished by the zebrafish *no tail* mutation. *Cell*, **75**, 99–111.

Horváth, L. & Orbán, L. (1995). Genome and gene manipulation in the common carp. *Aquaculture*, **129**, 157–82.

Inoue, K., Yamashita, S., Akita, N., Mitsuboshi, T., Nagahisa, E., Shiba, T. & Fujita, T. (1990). Electroporation as a new technique for producing transgenic fish. *Cellular and Differential Development*, **29**, 123–8.

Ivics, Z., Izsvák, Z. & Hackett, P.B. (1993). Enhanced incorporation of transgenic DNA into zebrafish chromosomes by a retroviral integration protein. *Molecular Marine Biology and Biotechnology*, **2**, 162–73.

Kalos, M. & Fournier, R.E.K. (1995). Position-independent transgene expression mediated by boundary elements from the apiloprotein B chromatin domain. *Molecular and Cellular Biology*, **15**, 198–207.

Kellum, R. & Schedl, P. (1992). A group of scs elements function as domain boundaries in an enhancer-blocking assay. *Molecular and Cellular Biology*, **12**, 2424–31.

Kimmel, C.B. & Warga, R.M. (1987). Cell lineages generating axial muscle in the zebrafish embryo. *Nature*, **327**, 234–7.

Kropp, K.E., Gulick, J. & Robbins, J. (1987). Structural and transcriptional analysis of a chicken myosin heavy chain gene subset. *Journal of Biological Chemistry*, **262**, 16536–45.

Lewis, E.B. (1950). The phenomenon of position effect. *Advances in Genetics*, **3**, 73–115.

Lin, S., Yang, S. & Hopkins, N. (1994*a*). *lacZ* expression in germline transgenic zebrafish can be detected in living embryos. *Developments in Biology*, **161**, 77–83.

Lin, S., Gaiano, N., Culp, P., Burns, J.C., Friedman, T., Yee, J.-K. & Hopkins, N. (1994*b*). Integration and germ-line transmission of a pseudotyped retroviral vector in zebrafish. *Science*, **265**, 666–9.

Liu, Z., Moav, B., Faras, A.J., Guise, K.S., Kapuscinski, A.R. & Hackett, P.B. (1990). Functional analysis of elements affecting expression of the α-actin gene of carp. *Molecular and Cellular Biology*, **10**, 3432–40.

MacGregor, G.R., Nolan, G.P., Fiering, S., Roederer, M. & Herzenberg, L.A. (1991). Use of *E.coli lacZ* β-galactosidase as a reporter gene. In *Methods in Molecular Biology*, 7, ed. E.J. Murray, pp. 217–235. Humana Press.

McKnight, R.A., Shamay, A., Sankaran, L., Wall, R.J. & Hennighausen, L. (1992). Matrix-attachment regions can impart position-independent regulation of a tissue-specific gene in transgenic mice. *Proceedings of the National Academy of Sciences, USA*, **89**, 6943–7.

Maclean, N., Penman, D. & Zhu, Z. (1987). Introduction of novel genes into fish. *Biotechnology*, **5**, 257–61.

Maclean, N. & Rahman, A. (1994). Transgenic fish. In *Animals with Novel Genes*, ed. N. Maclean, pp. 63–105. Cambridge University Press.

Marini, N.J., Etkin, L.D. & Benbow, R.M. (1988). Persistence and replication of plasmid DNA microinjected into early embryos of *Xenopus laevis*. *Developments in Biology*, **127**, 421–34.

Moav, B., Liu, Z., Caldovic, L.D., Gross, M.L., Faras, A.J. & Hackett, P.B. (1993). Regulation of expression of transgenes in developing fish. *Transgenic Research*, **2**, 153–61.

Mohun, T.J., Taylor, M.V., Garret, N. & Gurdon, J.V. (1989). The CArG promoter is necessary for muscle-specific transcription of the cardiac actin gene in *Xenopus* embryos. *EMBO Journal*, **8**, 1153–61.

Müller, F., Ivics, Z., Erdélyi, F., Papp, T., Váradi, L., Horváth, L., Maclean, N. & Orbán, L. (1992). Introducing foreign genes into fish eggs with electroporated sperm as a carrier. *Molecular and Marine Biology and Biotechnology*, **1**, 276–81.

Müller, F., Lele, Z., Váradi, L., Menczel, L. & Orbán, L. (1993). Efficient transient expression system based on square pulse electroporation and *in vivo* luciferase assay of fertilized fish eggs. *FEBS Letters*, **324**, 27–32.

Müller, H.P., Sogo, J.M. & Schaffner, W. (1989). An enhancer stimulates transcription in *trans* when attached to the promoter via protein bridge. *Cell*, **58**, 767–77.

Müller, H.P. & Schaffner, W. (1990). Transcriptional enhancers can act in *trans*. *Trends in Genetics*, **6**, 300–4.

Nilsson, E. & Lendahl, U. (1993). Transient expression of a human α-actin promoter/*lacZ* gene introduced into mouse embryos correlates with a low degree of methylation. *Molecular and Reproductive Development*, **34**, 149–57.

Ozato, K., Kondoh, H., Inohara, H., Iwamatsu, Y. & Okada, T.S. (1986). Production of transgenic fish: introduction and expression of chicken-crystallin gene in medaka embryos. *Cell Differentiation*, **19**, 237–44.

Petropoulos, J.C., Rosenberg, M.P., Jenkins, N.A., Copeland, N.G. & Hughes, S.H. (1989). The chicken skeletal muscle a-actin promoter is tissue specific in transgenic mice. *Molecular and Cellular Biology*, **9**, 3785–92.

Powers, D.A., Hereford, L., Cole, T., Creech, K., Chen, T.T., Lin, C.M., Kight, K. & Dunham, R. (1992). Electroporation: a method for transferring genes into the gametes of zebrafish (*Brachydanio rerio*), channel catfish (*Ictalurus punctatus*), and common carp (*Cyprinus carpio*). *Molecular and Marine Biology and Biotechnology*, **1**, 301–9.

Reinhardt, E., Nedivi, E., Wegner, J., Skene, J.H.P. & Westerfield, M. (1994). Neural selective activation and temporal regulation of a mammalian GAP-43 promoter in zebrafish. *Development*, **120**, 1767–75.

Rinder, H., Bayer, T.A., Gertzen, E.-V. & Hoffman, W. (1992). Molecular analysis of the ependymin gene and functional test of its promoter region by transient expression in *Brachydanio rerio*..*DNA and Cell Biology*, **11**, 425–32.

Rindt, H., Gulick, J., Knotts, S., Neumann, J. & Robbins, J. (1993). *In vivo* analysis of the murine β-myosin heavy chain gene promoter. *Journal of Biological Chemistry*, **268**, 5332–8.

Rossant, J. & Hopkins, N. (1992). Of fin and fur: mutational analysis of vertebrate embryonic development. *Genes Development*, **6**, 1–13.

Sands, A.T., Hansen, T.N., Demayo, F.J., Stanley, L.A., Xin, L. & Schwartz, R.J. (1993). Cytoplasmic β-actin promoter produces germ cell and preimplantation embryonic transgene expression. *Molecular and Reproductive Development*, **34**, 117–26.

Sekkali, B., Belayew, A., Martial, J.A., Hellemans, B.A., Ollevier, F. & Volckaert, F.A. (1994). A comparative study of reporter gene activities in fish cells and embryos. *Molecular and Marine Biology and Biotechnology*, **3**, 30–4.

Sin, F.Y.T., Bartley, A.L., Walker, S.P., Sin, I.L., Symonds, J.E., Hawke, L. & Hopkins, C.L. (1993). Gene transfer in chinook salmon (*Oncorchynchus tshawytscha*) by electroporating sperm in the presence of pRSV–*lacZ* DNA. *Aquaculture*, **117**, 57–69.

Sippel, A.E., Saueressig, H., Winter, D., Grewal, T., Faust, N., Hecht, A. & Bonifer, A. (1992). The regulatory domain organization of eukaryotic genomes: implications for stable gene transfer. In *Transgenic Animals*, ed. F. Grosveld & G. Kollias, pp. 1–26. Academic Press.

Skarnes, W.C. (1990). Entrapment vectors: a new tool for mammalian genetics. *Bio/Technology*, **8**, 827–31.

Stief, A., Winter, D.M., Stratling, W.H. & Sippel, A.E. (1989). A nuclear DNA attachment element mediates elevated and position-independent gene activity. *Nature*, **341**, 343–5.

Stuart, G.W., McMurray, J.V. & Westerfield, M. (1988). Replication, integration and stable germ-line transmission of foreign sequences injected into early zebrafish embryos. *Development*, **103**, 403–12.

Stuart, G.W., Vielkind, J.R., McMurray, J.V. & Westerfield, M. (1990). Stable lines of transgenic zebrafish exhibit reproducible patterns of transgene expression. *Development*, **109**, 577–84.

Subramaniam, A., Gulick, J., Neumann, J., Knotts, S. & Robbins, J. (1992). Transgenic analysis of the thyroid-responsive elements in the α-cardiac myosin heavy chain gene promoter. *Journal of Biological Chemistry*, **268**, 4331–6.

Szelei, J., Váradi, L., Müller, F., Erdélyi, F., Orbán, L., Horváth, L. & Duda, E. (1994). Liposome-mediated gene transfer in fish embryos. *Transgenic Research*, **3**, 116–19.

Tewari, R., Michard-Vanhee, C., Perrot, E. & Chourrout, D. (1992). Mendelian transmission, structure and expression of transgenes following their injection into the cytoplasm of trout eggs. *Transgenic Research*, **1**, 250–60.

Vielkind, J.R. (1992). Medaka and zebrafish: ideal as transient and stable transgenic systems. In *Transgenic Fish*, ed. C.L. Hew & G.L. Fletcher, pp. 72–91. World Scientific.

Volckaert, F.A., Hellemans, B.A., Galbusera, P., Sekkali, B., Belayew, A. & Ollevier, F. (1994). Replication, expression, and fate of foreign DNA during embryonic and larval development of the African catfish (*Clarias gariepinus*). *Molecular Marine Biology and Biotechnology*, **3**, 57–69.

Westerfield, M., Wegner, J., Jegalian, B.G., DeRobertis, E.M. & Püschel, A.W. (1992). Specific activation of mammalian Hox promoters in mosaic transgenic zebrafish. *Genes Development*, **6**, 591–8.

Weydert, A., Barton, P., Harris, A.J., Pinset, C. & Buckingham, M. (1987). Developmental pattern of mouse skeletal myosin heavy chain gene transcripts *in vivo* and *in vitro*. *Cell*, **49**, 121–9.

Zelenin, A.V., Alimov, A.A., Barmintzev, V.A., Beniumov, A.O., Zelenin, I.A., Krasnov, A.M. & Kolesnikov, V.A. (1991). The delivery of foreign genes into fertilized fish eggs using high-velocity microprojectiles. *FEBS Letters*, **287**, 118–20.

Zhu, Z., Li, G., He, L. & Chen, S. (1985). Novel gene transfer into the fertilized eggs of the goldfish. (*Carassius auratus* L. 1758). *Journal of Applied Ichthyology*, **1**, 31–4.

P. PRUNET, O. SANDRA and B. AUPERIN

Molecular characterization of prolactin receptor in tilapia

Introduction

Prolactin (PRL) is a polypeptide hormone, which in all vertebrates except Cyclostoms is synthesized in the adenohypophysis (Schriebman, 1986). PRL belongs to a family of structurally and functionally related hormones which include growth hormone, placenta lactogen, proliferin in mammals and somatolactin in fish. Based on amino acid sequence homologies, it has been reported that these proteins have arisen by duplication of an ancestral gene (Bewley & Li, 1971; Takayama *et al.*, 1991). Although in man, rat and turkey, PRL is the product of a single gene, the hormone occurs in multiple molecular forms, cleaved, phosphorylated or glycosylated whose discrete functions remain a matter of debate (Lewis *et al.*, 1984; Clapp *et al.*, 1989; Brooks *et al.*, 1990). Among teleost fish, PRL was isolated in the tilapia species *Oreochromis mossambicus* (Specker *et al.*, 1985), salmonids (Idler, Shamsuzzaman & Burton, 1978; Kawauchi *et al.*, 1983; Prunet & Houdebine, 1984; Anderson, Skibeli & Kautvik, 1989), carp (Yasuda *et al.*, 1987) and eel (Suzuki *et al.*, 1991). Amino acid sequence identity between fish and mammalian PRLs is only 20–30% whereas it increases to 60–80% between fish PRLs. The major difference between fish and mammalian PRLs is the absence of one disulfide loop in the *N*-terminal region. Nucleotide and polypeptide sequences analysis of these fish PRLs revealed the existence of two distinct, albeit similar, genes in some species. This is the case for salmon or carp (Yasuda, Itoh & Kawauchi, 1986, 1987).

In tilapia species, *Oreochromis mossambicus* and *O. niloticus*, a somewhat different situation was described, and two much less similar PRL molecules were characterized (Specker *et al.*, 1985; Yamaguchi *et al.*, 1988; Rentier-Delrue *et al.*, 1989). The larger PRL form (named $tiPRL_I$ or $tiPRL_{188}$) is a 20 836 Da protein containing 188 amino acids whereas the smaller form (named $tiPRL_{II}$ or $tiPRL_{177}$) is a 19 584 Da protein of 177 amino acids. The two forms have different isoelectric points (8.7 and 6.7 respectively) and the smaller form showed an

11-amino acid deletion. Comparison of the polypeptide sequences indicated that their identity was only 69%, the larger form ($tiPRL_I$) being more similar to other fish PRLs than to the smaller form ($tiPRL_{II}$). These analyses also clearly showed that the two PRL forms characterized in tilapia species are the product of distinct genes, the expression product of which has been recently localized in the same cells of the adenohypophysis (Specker *et al.*, 1993).

In fish, PRL exhibits a pleiotropic spectrum of biological activities including effects on osmoregulation, metabolism, behaviour and reproduction (for review see Clarke & Bern, 1980). However, the primary and most studied role of PRL in fish is regulation of hydromineral balance (Loretz & Bern, 1982; Hirano, 1986; Prunet *et al.*, 1990). Numerous studies on the osmoregulatory roles of PRL were performed in tilapia species and demonstrated that PRL is necessary for survival in hypo-osmotic environment (Dharmamba *et al.*, 1967; Dharmamba & Maetz, 1972). Moreover, in seawater-adapted tilapia, ovine PRL inhibited salt extrusion by gill chloride cells, thus inducing profound disturbance of hydromineral balance (Dharmamba & Maetz, 1976; Foskett, Machen & Bern, 1982). In this context, presence of two PRL forms in tilapia species raises interesting questions about their biological roles, especially in the control of osmoregulation.

Osmoregulatory functions of the two tilapia PRL forms

The osmoregulatory effects of the two tiPRLs forms have been tested in various biological assays. Thus, in freshwater-adapted tilapia *Oreochromis mossambicus*, both forms exhibited the same Na^+-retaining effect and the same inhibitory effect on transepithelial potential measurements in hypophysectomized fish (Specker *et al.*, 1985; Young *et al.*, 1988). Moreover, in intact tilapia, the two PRLs stimulated similarly plasma Ca^{2+} levels (Swennen *et al.*, 1991). However, Specker *et al.* (1993) have shown that the salamandrid integumental bioassay can distinguish between the two PRL forms: $tiPRL_{II}$ significantly decreased whole animal transepithelial potential whereas $tiPRL_I$ was without effect. The osmoregulatory role of these two hormones has been recently studied during adaptation to hyperosmotic environment (Auperin *et al.*, 1994*a*) using biologically active recombinant tiPRLs which allowed development of specific radioimmunoassays for both PRLs (Swennen *et al.*, 1991; Auperin *et al.*, 1994*a*). After direct transfer of tilapia *Oreochromis niloticus* to brackish water, plasma $tiPRL_I$ levels dropped rapidly below the detection threshold whereas measurable and significant levels of $tiPRL_{II}$ persisted. A similar pattern of change was observed when measuring pituitary contents of $tiPRL_I$ and $tiPRL_{II}$ after transfer to brackish water. These differences between the two forms were further confirmed by studying the effects of repeated injections of each PRL form in brackish water-adapted tilapia: $tiPRL_I$ induced clear and dose-dependent ion-retaining effects whereas $tiPRL_{II}$ produced markedly lower effects which were not dose related. These results clearly indicated that, in tilapia *Oreochromis niloticus*, the two PRL forms have different osmoregulatory roles during adaptation to hyperosmotic environment.

These findings in tilapia raise an interesting question on whether these osmoregulatory effects of PRL are associated with different PRL receptors. In higher vertebrates, the multiplicity of PRL actions appears to be reflected at the level of receptors for which multiple forms have been identified and suggest different signal transduction mechanisms (for review see Kelly *et al.*, 1991). A similar picture could also occur in fish, and leads us to question whether different receptor forms may be present in tilapia. In order to perform this characterization, studies of binding sites for PRL have been conducted in osmoregulatory organs. These studies were also completed by isolation of a cDNA encoding PRL receptor and analysis of its gene expression in these tissues.

Characterization of a high affinity binding site for PRL in tilapia gill and kidney

Because ovine PRL is probably unable to distinguish PRL from GH receptors in tilapia and thus leads to ambiguous interpretations (Prunet & Auperin, 1994), full characterization of PRL receptors in tilapia species requires the development of homologous radioreceptor assays using $tiPRL_I$ and $tiPRL_{II}$ as ligands. Such study was developed using recombinant tiPRL forms (Auperin *et al.*, 1994*b*). Measurements of the specific binding of ^{125}I-labelled $tiPRL_I$ and ^{125}I-labelled $tiPRL_{II}$ to membrane preparations from various tilapia tissues indicated that kidney and gill exhibited the highest levels of binding. Interestingly, the specific binding was always higher for $tiPRL_I$ than for $tiPRL_{II}$. The other tissues, including the liver, displayed significant but low specific binding (Prunet & Auperin, 1994).

Further studies of $tiPRL_I$ and $tiPRL_{II}$ binding sites (Auperin *et al.*, 1994*b*) were performed on gill and kidney from fish transferred for 36–48 h to brackish water: in this situation, an increase in specific binding for both tiPRL forms was observed. Scatchard analysis of $tiPRL_I$ and $tiPRL_{II}$ binding data suggested presence of a single class of high affinity binding sites recognized by both hormones. This result was observed for both kidney and gill membrane preparations. Interestingly, the affinity constants were always significantly higher for $tiPRL_I$ than for $tiPRL_{II}$. Specificity experiments were also conducted with various hormone preparations. In these studies, $tiPRL_I$ always appeared 8–10 times more potent than $tiPRL_{II}$ in competing for both gill and kidney binding sites, whichever tiPRL was used as a ligand. Furthermore, $tiPRL_I$ and $tiPRL_{II}$ totally displaced both tiPRL ligands from their gill and kidney binding sites, a result which further supports the presence of a unique PRL receptor in these organs. Overall, the above results revealed in gill and kidney tissues the presence of only one class of high affinity PRL receptors to which $tiPRL_I$ binds with a higher affinity than $tiPRL_{II}$.

Various growth hormone (GH) preparations were also tested for their ability to bind to tilapia PRL receptors (Auperin *et al.*, 1994*b*). Ovine GH and recombinant tiGH did not significantly displace any tiPRLs ligand from gill or kidney preparations. This indicates that, although PRLs and GHs have structural and evolutionary similarities, receptor characterized in this study is PRL specific. Thus, as competitor for the kidney PRL receptors, trout PRL appeared to be as effective as $tiPRL_I$, whereas carp PRL was less effective (Prunet & Auperin, 1994). Similarly, mammalian lactogenic hormones were able to bind to tilapia kidney PRL receptor but with less potency than any fish PRL.

The presence of a single tiPRL receptor class in gill and kidney, with higher affinity for $tiPRL_I$ than for $tiPRL_{II}$, does not explain the results obtained *in vivo* with brackish water-adapted tilapia treated with tiPRLs. However, we cannot be certain that our experimental conditions are adequate for characterizing another tiPRL receptor in osmoregulatory organs as Scatchard analysis would not necessarily reveal a second binding site. This would be the case particularly if the second receptor was low in concentration with binding affinity close to the values reported for the previous PRL receptor. Thus, in order to determine whether multiple forms of PRL receptor are present in tilapia osmoregulatory organs, cloning of a cDNA encoding tilapia PRL receptor was attempted.

Expression cloning of a tilapia PRL receptor

All classical attempts for cloning a fish PRL receptor cDNA using oligonucleotide probes or PCR strategy have been so far unsuccessful. However, binding results of tiPRL forms on gill and kidney membrane preparations previously described have allowed Sandra *et al.* (1995) to perform cloning this cDNA using an expression cloning strategy. This technique is based on the autoradiographic detection of the stable complex formed between a radioactive ligand and its specific receptor. A cDNA expression library constructed with poly(A)$^+$ RNA from kidney of freshwater-adapted tilapia was transiently transfected in COS cells and was screened with ^{125}I-tiPRL$_I$. This ligand was shown to present the highest specific binding and affinity constant for kidney PRL receptors (Auperin *et al.*, 1994*b*) which is probably important for the success of this expression cloning strategy.

This new cloning approach allowed the isolation of a ~3 kb clone. Analysis of the nucleotide and amino acid sequences confirmed similarities of this fish receptor with mammalian PRL receptors. The overall structure of the mature protein (606 amino acids) is similar to the long form of PRL receptor initially described in rat liver (Boutin *et al.*, 1988). Thus, some common features, previously described in all other PRL receptors isolated in higher vertebrates (Kelly *et al.*, 1991), are also present in the tilapia PRL receptor. As indicated in Fig. 1, this includes (i) an extracellular domain containing two pairs of cysteine residues and a Trp–Ser–Xaa–Trp–Ser motif but with no tandem repeat as reported in the avian PRL receptor (Tanaka *et al.*, 1992; Chen and Horseman, 1994), (ii) a short and unique transmembrane domain and (iii) a long cytoplasmic domain exhibiting the membrane proximal sequence, containing a proline-rich region and termed box 1, which is implicated in signal transduction (Lebrun *et al.*, 1994; Dusanter-Fourt *et al.*, 1994). However, the tilapia PRL receptor lacks a potential *N*-glycosylation site between the two pairs of cysteine residues and the overall amino acid identity with other PRL receptors is low (less than 40%). This gives clear evidence for the evolutionary changes undergone by the tiPRL receptor.

Expression of the tiPRL receptor in COS cells allowed characterization in these cells of a tiPRL$_I$ high affinity binding site. Analysis of tiPRL binding to the cell membrane preparations revealed differences in the biochemical characteristics of this receptor. When comparing with results obtained with kidney microsomes, the association constant obtained with COS cells is lower ($K_a = 1.7 \times 10^9$ M^{-1} and 2.9×10^{10} M^{-1}, respectively, for COS cells and kidney).

Moreover, as a competitor for tiPRL receptor in COS cells, ovine PRL was more potent than tiPRL$_{II}$, a result which is different to what has been described with kidney membrane preparations (Auperin *et al.*, 1994*b*). This is probably due to differences in membrane composition between mammalian COS cells and tilapia kidney cells.

Gene expression of PRL receptor in tilapia was further studied in fish reared in freshwater (Sandra *et al.*, 1995). Northern blot analysis indi-

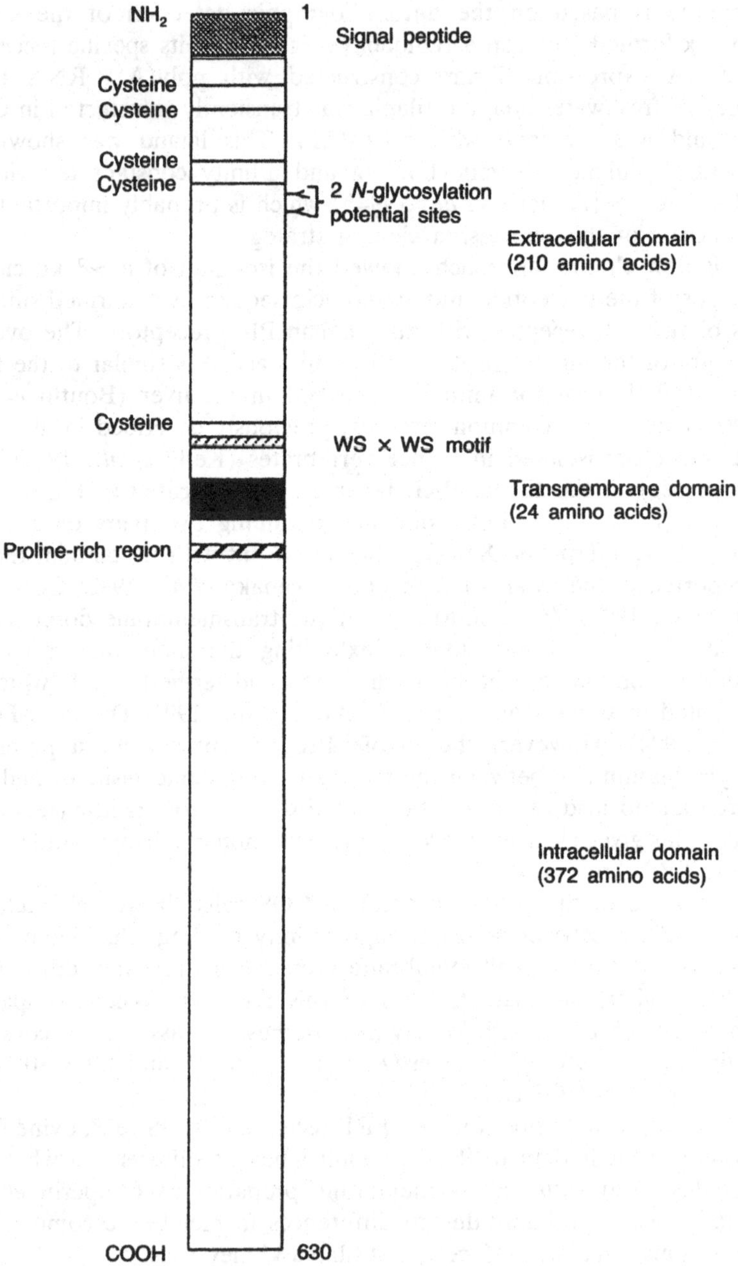

Overall structure of the tilapia prolactin receptor

cated the presence of a single transcript of ~3.2 kb in various tissues: a strong signal was observed in osmoregulatory organs (gill, kidney, intestine) whereas a clear but weak signal was seen in testis, skin and liver. No transcript was observed in muscle.

These results are in agreement with the binding studies presented above, and provide an additional argument in favour of a single high affinity receptor in osmoregulatory organs from tilapia.

Regulation of PRL receptor in gill tissue after a change of salinity

As indicated at the start of this chapter, differences in osmoregulatory functions between the two forms of tiPRL have been demonstrated in brackish water-adapted tilapia *Oreochromis niloticus* (Auperin *et al.*, 1994*a*). Thus, it appeared interesting to carry out a time-course study of PRL receptor changes during the freshwater–brackish water transfer at the level of gill tissue which is a major osmoregulatory organ during adaptation to a hyperosmotic environment.

After direct transfer of tilapia from freshwater to brackish water (22‰ salinity), an initial hydromineral imbalance was observed as early as 3 h after salinity changes, but, within 3 days, the fish appeared to adapt to hyperosmotic environment (Auperin *et al.*, 1994*a*). Specific binding of $tiPRL_{I}$ and $tiPRL_{II}$ was found to increase four to sevenfold and two to fivefold, respectively, within 24 h after transfer and the levels remained high until the end of the experiment (28 days in brackish water). Scatchard analysis, using $tiPRL_{I}$ as a tracer, indicated presence of only one class of PRL receptors in the gill of both freshwater and brackish water-adapted fish. In addition, $tiPRL_{I}$ always displayed a higher affinity than $tiPRL_{II}$ for tilapia gill receptors. The effect of salinity changes on PRL receptor transcripts in the gill of tilapia *Oreochromis niloticus* was also examined by Sandra *et al.* (1995) using cDNA probe corresponding to the extracellular domain of the receptor. Northern blot analysis of tiPRL receptor mRNA in gill showed a single transcript of ~3.2 kb both in freshwater-adapted tilapia and in fish transferred for 6 days to brackish water. Altogether, these

---

Fig. 1. Schematic representation of the tilapia PRL receptor identified in kidney. The transmembrane domain is shown in black. The WS × WS motif is a conserved tryptophan–serine–any residue–tryptophan–serine sequence which lies close to the transmembrane region.

results support the view that, during adaptation of tilapia to brackish water, both tilapia PRL forms bind to the same receptors in gill.

Time-course modifications of tilapia PRL receptor affinity and capacity were also analysed after transfer into brackish water by Auperin *et al.* (1995). As early as 24 h after transfer, the binding affinities significantly increased and remained high in tilapia adapted to hyperosmotic environment. Although the hypothesis of appearance of a new class of tilapia PRL receptors cannot be discarded, the authors rather suggested that the higher affinity was due to changes in the surface potential and ionization of the membrane (Hazel & Williams, 1991). This also could be due to the tiPRL receptor adopting a different conformation as a result of increased fluidity of gill membrane after the change in salinity (Leray *et al.*, 1984; Madsen, Meddings & Fedorak, 1992). Receptor capacity measured with the $tiPRL_I$ form increased sharply and considerably after transfer to brackish water (3.7 fold within 24 h), remaining high even when the fish were fully adapted to 22‰ salinity. Interestingly, this increase in number of free receptor was associated with a simultaneous drop in plasma $tiPRL_I$ and $tiPRL_{II}$ levels. By analogy to observations in mammals, this increase in PRL free binding sites could reflect a desaturation process of receptors previously occupied by endogenous hormone (Djiane, Delouis & Kelly, 1982). Such a hypothesis was further supported by quantification of PRL receptor transcript levels in gill. Transfer of tilapia to brackish water resulted in a low but significant decrease in tiPRL mRNA levels (Sandra *et al.*, 1995).

In conclusion, adaptation of tilapia *Oreochromis niloticus* to salinity is mainly associated with a drop in tiPRLs plasma levels presumably leading to the increase in the tiPRL-binding capacity of gill tissue. This result could also be discussed with regard to the environment in which tilapia live. Thus, Prunet and Auperin (1994) suggest that such an abundance of free PRL receptors in the gill of brackish water-adapted fish could be a means of adapting the hydromineral balance of the fish to a sudden drop in salinity.

Conclusions

All various approaches developed in order to characterize PRL receptors in osmoregulatory organs of tilapia *Oreochromis niloticus* support the presence of a single tiPRL receptor class to which either PRL forms bind. This conclusion does not really stand in line with the differences in osmoregulatory functions as discussed in the first paragraph. The observation that $tiPRL_I$ binds gill and kidney PRL receptors with higher affinity than tiPRL$_{II}$ is in agreement with the comparison of their biological potency in brackish water-adapted tilapia (Auperin *et al.*, 1994*a*). However, absence of dose-related effects of tiPRL$_{II}$ cannot be explained on the sole basis of lower affinity of this PRL form for the receptor in osmoregulatory organs. Thus, the most logical explanation at the present time would be to suggest that the two tiPRL forms act through different transduction mechanisms at the level of their common receptor. Indeed, there is now an abundant literature in higher vertebrates suggesting that different signal transduction pathways for PRL receptor may be involved. PRL receptor probably exists as a complex associating one molecule of ligand with two molecules of binding proteins (Fuh *et al.*, 1993) and several signalling molecules appear to play a role in signal transduction (for review see Horseman and Yu-Lee, 1994). This represents a whole area of research which needs to be developed in fish.

References

Anderson, O., Skibeli, V. & Kautvik, K.M. (1989). Purification and characterization of Atlantic salmon prolactin. *General Comparative Endocrinology*, **73**, 354–60.

Auperin, B., Rentier-Delrue, F., Martial, J.A. & Prunet, P. (1994*a*). Evidence that two tilapia (*Oreochromis niloticus*) prolactins have different osmoregulatory functions during adaptation to a hyperosmotic environment. *Journal of Molecular Endocrinology*, **12**, 13–24.

Auperin, B., Rentier-Delrue, F., Martial, J.A. & Prunet, P. (1994*b*). Characterization of a single prolactin receptor in tilapia (*Oreochromis niloticus*) which bind both tiPRL$_{I}$ and tiPRL$_{II}$. *Journal of Molecular Endocrinology*, **13**, 241–51.

Auperin, B., Rentier-Delrue, F., Martial, J.A. & Prunet, P. (1995). Regulation of gill prolactin receptors in tilapia (*Oreochromis niloticus*) after a change in salinity or hypophysectomy. *Journal of Endrocinology*, **145**, 213–20.

Bewley, T.A. & Li, C.H. (1971). Sequence comparison of human pituitary growth hormone, human chorionic somatomammotropin and ovine pituitary growth and lactogenic hormone. *Experimentia*, **27**, 1368–71.

Boutin, J.M., Jolicoeur, C., Okamura, J., Gagnon, J. Edery, M., Shirota, M., Danville, D., Fourt, I., Djiane, J. & Kelly, P.A. (1988). Cloning and expression of the rat PRL receptor, a member of the GH/PRL receptor gene family. *Cell*, **53**, 69–77.

Brooks, C.L., Kim, B.G., Aphale, P., Kleeman, B.E. & Johnson, G.C. (1990). Phosphorylated variant of bovine prolactin. *Molecular and Cellular Endocrinology*, **71**, 117–23.

Chen, X. & Horseman, N.D. (1994). Cloning, expression, and mutational analysis of the pigeon prolactin receptor. *Endocrinology*, **135**, 269–76.

Clapp, C., Sears, P.S., Russel, D.H., Richards, J., Levay-Young, B.K. & Nicoll, C.S. (1989). Biological and immunological characterization of cleaved 16K forms of rat prolactin. *Endocrinology*, **122**, 2892–8.

Clarke, W.C. & Bern, H.A. (1980). Comparative endocrinology of prolactin. In *Hormonal Proteins and Peptides*, ed. C.H. Li, vol. 8, pp. 105–197. New York: Academic Press.

Dharmamba, M. & Maetz, J. (1972). Effects of hypothysectomy and prolactin on the sodium balance of *Tilapia mossambica* in fresh water. *General and Comparative Endocrinology*, **19**, 175–83.

Dharmamba, M. & Maetz, J. (1976). Branchial sodium exchange in seawater-adapted *Tilapia mossambica*. Effects of prolactin and hypophysectomy. *Journal of Endocrinology*, **70**, 293–9.

Dharmamba, M., Handin, R.I., Nandi, J. & Bern, H.A. (1967). Effect of prolactin on freshwater survival and on plasma osmotic pressure of hypophysectomized *Tilapia mossambica. General and Comparative Endocrinology*, **6**, 295–302.

Djiane, J., Delouis, C. & Kelly, P.A. (1982). Prolactin receptor turnover in pseudopregnant rabbit mammary glands. *Molecular and Cellular Endocrinology*, **25**, 163–70.

Dusanter-Fourt, I., Muller, O., Ziemiecki, A., Mayeux, P., Drucker, B., Djiane, J., Wilks, A., Harpur, A.G., Fischer, S. & Gisselbrecht, S. (1994). Identification of JAK protein tyrosine kinases as signalling molecules for prolactin. Functional analysis of prolactin receptor and prolactin-erythropoietin receptor chimera expressed in lymphoid cells. *EMBO Journal*, **13**, 2583–91.

Foskett, J.K., Machen, T.E. & Bern, H.A. (1982). Chloride secretion and conductance of teleost opercular membrane: effects of prolactin. *American Journal of Physiology*, **242** R380–9.

Fuh, G., Cotosi, P., Wood, W.I. & Wells, J.A. (1993). Mechanism-based design of prolactin receptor antagonists. *Journal of Biological Chemistry*, **268**, 5376–81.

Hazel, J.R. & Williams, E.E. (1991). The role of alterations in membrane lipid composition in enabling physiological adaptation of organisms to their physical environment. *Progress in Lipid Research*, **29**, 167–227.

Hirano, T. (1986). The spectrum of prolactin action in teleosts. In *Comparative Endocrinology: Developments and Directions*, ed. C.L. Ralph, pp. 53–74. New York: Liss.

Horseman, N.D. & Yu-Lee, L.Y. (1994). Transcriptional regulation by the helix bundle peptide hormones: growth hormone, prolactin, and hematopoietic cytokines. *Endocrine Review*, **15**, 627–49.

Idler, D.R., Shamsuzzaman, K.M. & Burton, M.P. (1978). Isolation of prolactin from salmon pituitary. *General and Comparative Endocrinology*, **35**, 409–18.

Kawauchi, H., Abe, K.I., Takahashi, A., Hirano, T., Hasegawa, S., Naito, N. & Nakai, Y. (1983). Isolation and properties of chum salmon prolactin. *General and Comparative Endocrinology*, **49**, 446–58.

Kelly, P.A., Djiane, J., Postel-Vinay, M.C. & Edery, M. (1991). The prolactin/growth hormone receptor family. *Endocrine Reviews*, **12**, 235–51.

Lebrun, J.J., Ali, S., Sofer, L., Ullrich, A. & Kelly, P.A. (1994). Prolactin-induced proliferation of Nb2 cells involves tyrosine phosphorylation of the prolactin receptor and its associated tyrosine kinase JAK2. *Journal of Biological Chemistry*, **269**, 1–6.

Leray, C., Chapelle, S., Duportail, G. & Florentz (1984). Changes in fluidity and 22:6 (n-3) content in phospholipids of trout intestinal brush-border membrane as related to environmental salinity. *Biochemica et Biophysica Acta*, **778**, 223–8.

Lewis, U.J., Singh, R.N.P., Lewis, U.J., Seavey, B.K. & Sinha, Y.N. (1984). Glycosylated ovine prolactin. *Proceedings of the National Academy of Sciences, USA*, **81**, 385–9.

Loretz, C.A. & Bern, H.A. (1982). Prolactin and osmoregulation in vertebrates. *Neuroendocrinology*, **35**, 292–304.

Madsen, K.L., Meddings, J.B. & Fedorak, R.N. (1992). Basolateral membrane lipid dynamics after Na^+–K^+-ATPase activity in rabbit small intestine. *Canadian Journal of Physiology and Pharmacology*, **11**, 1483–90.

Prunet, P. & Auperin, B. (1994). Prolactin receptors. In *Fish Physiology, Molecular Aspects of Hormonal Regulation in Fish*, ed. N. Sherwood & C.L. Hew, vol. XIII, pp. 367–391. San Diego: Academic Press.

Prunet, P. & Houdebine, L.M. (1984). Purification and biological characterization of chinook salmon prolactin. *General and Comparative Endocrinology*, **53**, 49–57.

Prunet, P., Avella, M., Fostier, A., Björnsson, B., Th., Boeuf, G. & Haux, C. (1990). Roles of prolactin in salmonids. In *Progress in Comparative Endocrinology*, ed. A. Epple, C.G. Scanes & M.T. Stetson, pp. 547–552. New York: Wiley-Liss.

Rentier-Delrue, F., Swennen, D., Prunet, P., Lion, M. & Martial, J.A. (1989). Tilapia prolactin: molecular cloning of two cDNAs and expression in *Escherichia coli*. *DNA*, **8**, 261–70.

Sandra, O., Sohm. F., de Luze, A., Prunet, P., Edery, M. & Kelly, P.A. (1995). Expression cloning of a cDNA encoding a fish prolactin receptor. *Proceedings of the National Academy of Sciences, USA*, **92**, 6037–41.

Schreibman, M.P. (1986). Pituitary gland. In *Vertebrate Endocrinology: Fundamentals and Biomedical Implications*, ed. P.K.T. Pang & M.P. Schriebman, vol. 1, pp. 11–55. Toronto: Academic Press.

Specker, J.L., King, D.S., Nishioka, R.S., Shirahata, K., Yamaguchi, K. & Bern, H.A. (1985). Isolation and partial characterization of a pair of prolactins released *in vitro* by the pituitary of a cichlid fish, *Oreochromis mossambicus. Proceedings of the National Academy of Sciences, USA*, **82**, 7490–4.

Specker, J.L., Kishida, M., Huang, L., King, D.S., Nagahama, Y., Ueda, H. & Anderson, T.R. (1993). Immunocytochemical immunogold localization of two prolactin isoforms in the same pituitary cells and in the same granules in the tilapia (*Oreochromis mossambicus*). *General and Comparative Endocrinology*, **89**, 28–38.

Suzuki, R., Yasuda, A., Kondo, J., Kawauchi, H. & Hirano, T. (1991). Isolation and characterization of Japanese eel prolactins. *General and Comparative Endocrinology*, **81**, 391–402.

Swennen, D., Rentier-Delrue, R., Auperin, B., Prunet, P., Flik, G., Wendellaar Bonga, S.E., Lion, M. & Martial, J.A. (1991). Production and purification of biologically active recombinant tilapia (*Oreochromis niloticus*) prolactins. *Journal of Endocrinology*, **131**, 219–27.

Takayama, Y., Ono, M., Rand-Weaver, M. & Kawauchi, H. (1991). Greater conservation of somatolactin a presumed pituitary hormone/prolactin family, than of growth hormone in teleost fish. *General and Comparative Endocrinology*, **83**, 366–74.

Tanaka, M., Maeda, K., Okubo, T. & Nakashima, K. (1992). Double antenna structure of chicken prolactin receptor deduced from the cDNA sequence. *Biochemical and Biophysical Research Communications*, **188**, 490–6.

Yamaguchi, K., Specker, J.L., King, D.S., Yokoo, Y., Nishioka, R.S., Hirano, T. & Bern, H.A. (1988). Complete amino acid sequences of pair of fish (tilapia) prolactins $tPRL_{177}$ and $tPRL_{188}$. *Journal of Biological Chemistry*, **263**, 9113–21.

Yasuda, A., Itoh, H. & Kawauchi, H. (1986). Primary structure of chum salmon prolactins: occurrence of highly conserved regions. *Archives in Biochemistry and Biophysiology*, **244**, 528–41.

Yasuda, A., Miyazima, K.I., Kawauchi, H., Peter, R.E., Lin, H.R., Yamaguchi, K. & Sano, H. (1987). Primary structure of common carp prolactins. *General and Comparative Endocrinology*, **66** 280–90.

Young, P.S., McCormick, S.D., Demarest, J.R., Lin, R.J., Nishioka, R.S. & Bern, H.A. (1988). Effects of salinity, hypophysectomy and prolactin on whole animal transepithelial potential in the tilapia. *Oreochromis mossambicus. General and Comparative Endocrinology*, **71**, 389–97.

Index

actin 95–6, 104–6
AF *see* antifreeze proteins
Antarctic cod 2, 4–13, 14–15
Antarctic fish
  antifreeze proteins 1–16
  muscle activity 76–7, 78
  thermal range 75–6
antibody injection 116–17
antifreeze proteins 1–16
  Antarctic cod, 2, 4–13, 14–15
  function 1
  gene coding 8–13, 15–16
  glycopeptides 2, 4–13, 15–16
  molecular structure 2–3, 4–7
  peptides 2–3
  polyprotein genes 8–13, 15–16
  types 2–3

blood salt 1

Ca-ATPase 60, 101–2
carp
  β-actin gene 166–7, 169–70, 171, 172
  muscle activity 77
  myosin heavy chain genes 78–9. 84–5, 124–43
  temperature adaptation 28–35, 77, 138–9
  transient expression 182–93
catfish, transient expression 182–93
CFTR 48, 59–60
CHH 95, 102
chloride cells 44–50
chloride channels 48, 59–60
CIP 36
cod, Antarctic 2, 4–13, 14–15
cold adaptation
  antifreeze proteins 1–16
  carp 28–35, 77, 138–9
  gene coding 8–13, 15–16, 27–36, 79
  membrane changes 21–38
  muscle performance 76–7
cold-inducible promoter 36

cortisol 49, 53
crustacean growth 93–108
crustacean hyperglycaemic hormone 95, 102
cystic fibrosis transmembrane conductance regulator 48, 59–60

Δ^9-desaturase 23, 24–5, 27–35
desaturases 21–38
  carp 28–35
  cold adaptation 23–38
  gene transcription 27, 30–6
  homeoviscous adaptation 23, 35–7
  induction 25–8, 35–7
  types 24
drinking reflex 43–4, 50

ecdysteroids 106–7
eel, European 43–4
enhancer effects 187–9, 190
enhancer trapping 178

gene trap 115–16
gill
  chloride cells 44–50
  prolactin receptors 203–4, 207–8
growth, crustacean 93–108
growth hormone 159–61

homeoviscous adaptation 21, 22, 23, 35–7
hormones *see also* prolactin
  growth 159–61
  ion transport 48–50, 53–4
  moulting 93–5, 102, 106–8

insertional mutagenesis 115
intestine
  antifreeze proteins 8
  ion transport 44, 50–4
ion cotransporters 48, 52–3
ion transport 43–61
  chloride cells 44–50

ion transport (*cont.*)
gene coding 54–60
hormonal regulation 48–50, 53–4
intestine 44, 50–4
modelling 45–7
Na, K-ATPase 47, 49, 50, 51–2, 53, 54–8

kidney, prolactin receptors 203–4

lacZ 166–72, 177, 181, 183–93

MIH 95, 102
modelling
ion transport 45–7
regulatory sequences 165, 175–6
using zebrafish 113–20, 175–6
vertebrate development 113–14
moult-inhibiting hormone 95, 102
moulting 93–5, 103–8
muscle *see also* myosin heavy chains
moulting 93–5, 103–4, 106–8
red and white 77, 133–8, 152
temperature changes 76–87, 138–9
myosin ATPase 76–8
myosin heavy chain genes 78–87, 96–100, 123–5, 150
carp 78–9, 84–5, 124–43
growth hormone 159–61
mammals 82–4, 131–2
polymorphism 154–8
transgenesis 181–93
myosin heavy chains 78, 123, 149 *see also* myosin heavy chain genes
growth hormone 159–61
polymorphism 150, 151–8
rainbow trout 152–8
myosin light chains 78, 81–2, 123, 149, 150
enhancer 187

Na,K-ATPase 47, 51–2
gene coding 54–8
hormonal regulation 49, 50, 53
Na/K/2Cl-contransporter 48, 52–3
gene coding 58–9
NC-AFPG gene 8–10, 11, 12
neurohormones 102

oesophagus, ion transport 50–4
opercular chloride cells 44–50
Oregon School 114
osmoregulation 43–4 *see also* ion transport
hormones 49, 202–3, 207–9

polymorphism 150, 151–8
polyprotein genes 8–13, 15–16
prolactin (PRL) 49, 201–2
osmoregulatory role 49, 202–3
receptors 203–9
tiPRLs 201–9
promoter efficiency 180–1, 182–93
protein antifreezes *see* antifreeze proteins

rainbow trout
growth hormone 159–61
myosin heavy chains 152–8
transgene expression 165–72
red muscle 77, 133–6, 152
regulatory sequences 165–72, 175, 180–1
regulon 178
reporter genes 165–72, 176–7 *see also* *lacZ*
RNA injection 117–20

salt, blood 1
sarco/endoplasmic reticulum Ca-ATPase 101–2
seawater adaptation 43–61, 207–8
SERCA 101–2
sodium pump *see* Na,K-ATPase

temperature adaptation *see also* cold adaptation
muscle function 75–87, 138–9
tilapia
osmoregulation 202–3, 207–8
prolactin receptors 203–9
transgene expression 165–72
tiPRLs
molecular structure, 201–2
receptors 203–9
transcription
desaturase activity 27, 30–6
myosin heavy chains 82–4
transgenesis 165, 171, 175–8
muscle tissue 181–93
zebrafish 115–16
transient expression 165–72, 177, 179–93
tropomyosin 100–1
trout *see* rainbow trout

vertebrate models 113–14, 175–6
viscotropic regulation 26

white muscle 136–8

zebrafish
models for vertebrates 113–20, 175–6
transient expression 180–1, 182–93

Printed in the United States
By Bookmasters